Lean Management 50-50-20©™

By

Accuride Corporation

Jd Marhevko

Arvind Srivastava

Mary Blair

Accuride Corporation

Evansville, IN

Lean@AccurideCorp.com

Editors: Alexa Druckmiller, Chad Monroe
Cover Design: Amy Pfaffmann
ISBN: 978-0-692-78836-3

Copyright 2016 by Accuride Corporation. First Edition.
All rights reserved. No part of this book may be reproduced or utilized in any form or by any means, electronic or mechanical, including photocopying, recording, or any information storage and retrieval system, without permission from Accuride.

Manufactured in the United States of America.

Dedication

This book is dedicated to the perpetual efforts of Accuride's personnel that have been a part of this journey as well as those who left Accuride and shared these lessons with others. More specifically, this goes to our Lean, Quality, Supply Chain, Operations, IT, HR, Finance, Sales and Legal team members that make up Accuride's integrated Quality/Lean Management System (QLMS) Council. This team plans, develops and executes Accuride's Lean system enterprise-wide. We can only do this with an interactive and engaged group of personnel. Businesses are in a constant flux when in pursuit of survival and growth. This Accuride council has leveraged Lean to help it do both. We need to position ourselves to be the supplier of choice. Not just for our high quality and cost effective products, but also for the value proposition that we can bring to our customers. This book is a dedicated reflection of their collective efforts; past, present and future. It is also dedicated to you, who we hope will benefit from what we have learned on our continued Lean journey.

Acknowledgment

The authors would like to thank Rick Dauch, Accuride President and CEO, for encouraging them to document and share Accuride's award winning Lean management system so that others could benefit from it as well. We appreciate Accuride's executive committee, specifically Scott Hazlett, Greg Risch, Paul Wright, Steve Martin and Mike Hajost, for spreading the Lean message deep and wide with their full support to make Accuride's Lean journey successful. The site-based cross functional teams are led by Accuride's Directors of Operations, a special breed unto themselves, in order to enable these systems. We include Operations Directors Mike Waller, Steve Kuhn, Greg Dauer, Eric Pansegrau, Kris Elliot, Chris Cain and Tim Rogers to this list of leaders. At our sites, we thank all of our Council members that make these things happen from day to day, hour to hour and minute to minute. To support a book in a Lean environment, one requires creativity. The authors are thankful for the enthusiastic support received from the overall Accuride team. We would like to specially thank our internal book reviewing team, Chad Monroe, Greg Risch and Timothy Weir. We'd like to thank Amy Pfaffmann and Ashley Steele for the front and back cover design.

We're deeply appreciative to several external subject matter experts for their critical feedback and reviews, including Dr. Bruce DeRuntz (SIU), JR McGee (X-Stream Leadership Group), Frank Murdock (FKM Consulting) and Kush Shah (ASQ Fellow, Past Chair Automotive Division). We'd also like to thank Harris Lean Systems for their support on our Lean journey. The authors are deeply thankful to Alexa Druckmiller for her thorough edits and to Mark Druckmiller for the logistics support to help with the technical execution of this book. Last, but not least, the authors would like to thank their families for their continued love, encouragement, trust and support.

Foreword

My first introduction to Lean systems came in the fall of 1990 while a student at MIT's Leaders for Manufacturing program. As part of our Operations Management class we read James Womack's book *The Machine that Changed the World*. In this classic book, Womack compared the management styles and operating and financial performance of the North American and Japanese car companies. While GM, Ford and Chrysler continued to lose market share and struggled with financial viability, the Japanese OEMs flush with cash (especially Toyota) were flourishing, growing globally and gaining market share with new, fresh and expanding product line ups. Why? The answer was their adoption and relentless implementation of Lean systems.

Fast forward 26 years, across four companies and countless Kaizen, Demand Flow, Six Sigma and other Lean training classes/seminars, and I am still learning how to be a true Lean practitioner. The quest to "get Lean" and remain competitive in the face of increased global competition continues. My business experiences have taken me to every major industrial country and region in the world. I have seen firsthand the differences between world class Lean companies and others. While many companies claim to operate Lean systems, the ugly fact is that only 10-15 percent *are truly* Lean. Lean companies tend to be the market leaders with superior operating and financial results. They use the cash generated by being Lean to invest back into their business – advanced R&D, new product development, capital investments to expand capacity and global footprint, strategic acquisitions – to further their profitable growth. Getting Lean means making money!

50-50-20™ was written to demonstrate how the adoption of Lean systems and tools drives bottom line results. We have taken Lean tools and applied them across our entire business – on the shop floor, in our distribution centers, in our back and front offices, in our IT systems, across our supplier partners and even to our customers receiving docks. The results speak for themselves – world class quality performance in COPE, warranty and customer ppm; 99-100 percent on-time delivery performance with 50 percent less inventory and dramatically reduced lead times; >20 percent productivity improvement across the enterprise. All of this adds up to improved free cash flow, EBITDA and earnings per share. Getting Lean means making money, which Eli Goldratt pointed out in his book *The Goal*.

Getting and being Lean is not a slogan or a management fad. It is a way of doing business and it is hard work. It takes the involvement of hourly and salaried teams across all functional disciplines and tremendous support from the C-Suite to make it happen. We hope that 50-50-20™ helps you and your team better understand the true power of being Lean. Good luck on your journey!

Rick Dauch, Accuride President and CEO

Lean Management 50-50-20™
Table of Contents

Introduction

There are hundreds of books written on Lean tools and techniques. From a Lean perspective, adding one more to the list is *Muda*. This book, therefore, does not try to explain ***how*** to use Lean tools but instead ***demonstrates*** their successful application across products, processes and partners as experienced by Accuride Corporation. This book is technically advanced and details the journey of a manufacturing organization. It is for Lean Professionals and Senior Leaders who have had exposure to Lean systems and its tools, techniques and terminology.

We are encouraged by the recognition of four of our facilities in three years by the Association for Manufacturing Excellence (AME). This recognition validated the breadth and depth of our Lean efforts and motivated us to share the results of our Lean journey with you. Across this journey, we have experienced that with an averaged 50% reduction in lead time that our teams were able to increase productivity by an average of 50% as well as reduce cost per unit by an average of 20%, hence the 50-50-20™ in the title. Each chapter of the book shares examples from Accuride's experience to illustrate how we used and benefited from it.

Peter Drucker once said, "In a bottle the neck is at the top, and that is the bottleneck." Any organizational Lean effort is successful only with top leadership support with leadership both in the C-suite and across the business. We need to deliver products and services effectively (best quality) and efficiently (with optimal resources). The role of leadership and quality is explained in Chapter 1. Once leadership roles are understood, Chapter 2 shows how to translate the vision, mission and objectives into actionable items. Chapter 3 expands on Chapter 2 via the review of leading and lagging metrics. Improved results are the outcome of focused actions. These actions can be identified during the mapping of the process with material and information data called Value Stream Mapping (VSM). Chapter 4 reviews various types of VSMs used to identify and address opportunities. Once the improvements are made, sustained results are described by a Visual Operation System (VOS), detailed in Chapter 5.

The full value of Lean cannot be realized by looking only within the four walls of a facility. The scope needs to be expanded to include both the enterprise and external partners. The concept of an Enterprise-VSM (E-VSM) is detailed in Chapter 6. Chapter 7 describes the corporate Lean system used to sustain momentum and monitor performance. Chapter 8 describes supply chain partnerships by extending the scope of the Lean systems. And finally, all Lean actions need to result in the outcomes necessary for a business to succeed, which are detailed in Chapter 9.

We hope that you will receive as much value as we have on this journey. Accuride would welcome your questions/comments/feedback at **Lean@AccurideCorp.com**.

Jd Marhevko, Arvind Srivastava and Mary Blair

Chapter One: Lean Leadership

50-50-20… That's a pretty powerful impact. To assure the reader that this is not just Accuride Corporation's word on these results, four of Accuride's North American manufacturing plants have been honored recipients of the Association of Manufacturing Excellence's (AME) Manufacturing Excellence Award over three years: Henderson, KY (2014); Erie, PA and Rockford, IL (2015); and Monterrey, Mexico (2016). Each site won on its first submission of the application. The caliber of the Lean auditors and the robustness of the evaluation process have strong similarities to those of the Performance Excellence (Baldrige) and Shingo Prize systems.

Common sense tells us that leadership is key to the success of any Lean implementation. For that common sense to be leveraged, this means that the leadership team needs to understand the power that Lean can have to make a difference in their organization. Accuride's leadership has not only drunk the "Lean Kool-Aid," they swam in it; and are swimming still. The book title reflects Accuride's experience of "50-50-20"™ across its first five years of its lean journey. Overall, Accuride was able to reduce the average Lead Time **(LT)** of more than eight production and 11 transactional processes by more than **50%** enabling an average **Productivity** increase of **50%** across the organization and an average **20%** reduction in controllable Cost Per Unit (**CPU**).

Accuride started with some basic assumptions:

1. The leadership had to be willing to learn, listen and commit itself to make improvements. Those that didn't have the aptitude and attitude are no longer with the organization.
2. We had to recover baseline product and service quality levels to align with our customers' changing expectations.
3. We needed to improve both our customer and supplier relationships to restore their trust in Accuride.
4. We needed to evolve from just making a "good product" to providing the best quality, highest reliability and on-time products "and services" at a competitive cost; enabling Accuride to provide additional value outside of "just a part."

Figure 1.1 shows part of the concept of the "50-50-20"©™ model. A 50% reduction in the lead time was enabled by executing various Lean activities such as reducing the process variation, balancing workloads, managing changeover and executing inventory kanban. Each of these actions reduced cost and improved product/process velocity and productivity. A key stumbling block is that if the product/information is not right to begin with, then efficient pull cannot occur. Tools like Design For Six Sigma (DFSS) and Six Sigma reduce process variation (Yes, they are different: DFSS is preventive and Six Sigma is typically reactive/after the fact). Velocity and pull is achieved via the use of Lean tools. This book is Accuride's primary approach to Lean. We define it as a set of concepts, tools and management processes aimed at strengthening a business' competitive advantage realized from its processes.

In this chapter we focus on leadership and product quality. We start with leadership because with the right leadership, the system falls in place via the establishment of the right processes to deliver sustained excellence in products, service and customer value.

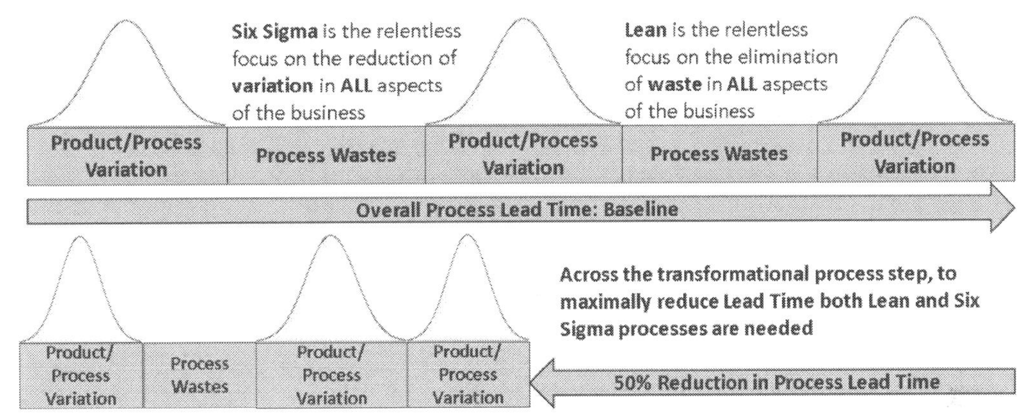

Figure 1.1

Leadership

Leaders inspire actions and support their teams to accomplish goals. A common perception of leadership is that it resides at the C-Suite. This part of the leadership team needs to develop the strategies, align the actions of the team to meet them and secure the necessary resources. While they're doing this, they need to lead by example. Their behaviors demonstrate what generally gets replicated. However, outside of the C-Suite, effective leadership is needed at all levels.

A successful execution of strategy needs to be internalized by the personnel performing the processes and executing process leadership. As Accuride recognized its varying types of leadership across a multi-site, multi-division organization, the strategic concepts needed to be interpreted broadly with the key intent of providing continuity. The leadership discussion is intentionally built from a servant leadership perspective from the process level through to the governance level. **Figure 1.2** shows leadership at different levels.

Figure 1.2

1. Process Leadership

What matters most for an execution of strategy are the actions and the decisions that take place where the actual value add occurs. The actions and the decisions of these personnel on the place of work have the most direct impact on the organization's actual operational and financial performance. Process leaders have the most frequent interaction with those adding product value. As such, the leadership styles, attitudes and support structures are critical organizational success factors.

In addition to their management/administrative roles, process leaders need to ***inspire*** and ***empower*** their teams to sustain focus on achieving team objectives. Demonstrating this type of interactive leadership for making continual improvements goes beyond originally planned objectives; this is culture. The strongest Lean organizations have some form of cultural development system in place. You feel it when you enter the facility. There is a pulse when you walk through an OC Tanner or Autoliv facility. At Accuride, we're continuing to build on our "AccuPride"™. There are dozens of "isles of excellence" that have positively overlapped into other areas. This outward spread happens when the positive, latent energy that is inherent in these teams can be utilized. This untapped energy is waiting to be released. The leadership team must simply be empowered to listen and act upon the feedback provided.

An example of an informal network of cultural flow is MOUND. We heard about this through the University of Tennessee's Lean program. MOUND stands for Middle, Out, Up aNd Down. Somewhere in the *middle* of the organization, an idea or improvement is successfully implemented. A peer leverages it *outward* to improve their area. This gets the attention of *upper* management who then furthers it horizontally *and down* across the organization. Organizations that need too many layers of permission can easily stifle MOUND. At Accuride, MOUND is alive and well.

We heard at an AME 2015 Conference Panel Discussion that ***change moves at the speed of trust***. This is a reflective concept that requires honest self-evaluation. Do we trust our team members to make the right decision and inform us later? Do they trust that we will not overcorrect or be punitive if an honest, proactive action goes awry? There are, of course, levels here that need to be respected in the safety of personnel and in the production of safety products. However, significant leeway can be positively enabled with amazing results.

The good news is that leadership can be learned, and behaviors can be monitored to ensure this is happening. This is where the broadest level of leadership (Facility) comes into play.

2. Facility Leadership

Facility-level leadership has the broadest reach across the business. Its role is to integrate and harmonize leadership among the various functions within the four "walls" of the immediate work cell, product/process stream or facility. They also act as an intermediary between the next level of business leadership sharing performance information and process support needs.

A critical leadership feature at this level is one of availability to the local team members to assess live performance status and to respond to team needs. Visual Operating Systems (VOS) are leveraged to communicate team and/or individual performance status. There are more than 20 different types of VOS utilized across Accuride. These create an instant visual status of the process at hand. One of Accuride's Director of Operations stated, "We manage by roaming around." Facility Directors typically spend more than 50 percent of their time in the work environment (or Gemba). They rely on employee interactions and the VOS to rapidly identify gaps or areas where processes are not performing as intended.

Another key form of communication is a Town Hall style meeting to help the overall site team understand the status of the business and to share directional information. In general, the communication styles used need to create an environment of cooperation and support. There needs to be a deliberate process in place to make it happen.

3. Division/Business Leadership

Division and/or business leadership, along with the functional support team leaders, such as Human Resources, Engineering, Supply Chain, Quality Management, Sales/Marketing, IT, Legal, etc., has a broader view of the organization. They work both vertically within their function and horizontally across the business to accomplish shared objectives. These leaders are typically removed one or two levels away from day to day operations. They rely on Key Performance Indicators (KPIs) to detect process gaps and performance trends. The leadership at this level:

a. Inspires teams to develop deeper functional knowledge for achieving performance objectives and managing cause and effect relationships.
b. Provides a non-threatening environment in which teams can uncover underlying issues so that they can be addressed adequately.
c. Educates teams so that they can evaluate the intended and unintended consequences of their actions.

d. Educates themselves so that they can continue to provide an optimized knowledge base for improving the business and providing effective input to the business strategy.
e. Emphasizes the relentless pursuit of improvement, which translates into external and internal customer satisfaction. Improvement should occur at two core levels:
    i. Improving the **efficacy** of processes: What are the key deliverables of this team? What are the key processes that enable those deliverables? How are they monitored for correctness/accuracy? Assuming the levels are not at 100%, how are the processes being improved upon? Are the processes really needed, or are they a form of "inspection or rework" that can be eliminated?
    ii. Improving the **efficiency** of processes: Based on the identified processes that are considered core to the function, how has it been determined that these are being performed with the least amount of work time?
    iii. When evaluating transactional efficacy and efficiency, Transactional-Value Stream Maps (T-VSM) are used to demonstrate the process flow, understand the inputs and outputs and review the system overall for improvement (See Chapter 4).

4. Organizational Leadership

A Chief Executive Officer (CEO) generally provides organizational leadership in which he or she develops and enables the division/business unit leadership. This team translates and provides analysis and inputs for developing and managing the business strategy. They monitor KPIs and enable support and gather resources to execute the plan. A clear line of deployment is needed both horizontally across the organization as well as vertically into the various layers of the organization. A common tool used to deploy the strategies within several of the functions is called Hoshin Kanri. Accuride calls it the "X-Matrix" (See Chapter 2). Accuride has a strong percentage of personnel with military experience in its organizational leadership team. Integrity and honor are an unwritten and expected code of performance. This level of leadership is the face-to-face conduit to other stakeholders, such as the shareholders, the community at large, government, media, legal authorities, etc. Leadership at this level acts as the head coach to the organization in developing and promoting a sustainable leadership at all the levels.

5. Board Governance

The board consists either of Owners (if privately held) or a Board of Director's (BOD) (if publicly held). This team needs to ensure that:

a. Organizational activities are aligned to the long-term interests of the stockholders.
b. A robust system is in place to adequately manage risk.

The BOD has the ultimate decision-making authority and assumes the legal responsibility for corporate activities. This team ensures that the Organizational Leadership is effective and able to accomplish the organizational vision, mission and objectives.

Quality

The customer defines quality. Per Juran's *Quality Control Handbook*, two meanings dominate in the use of the word "Quality:"

1. Product features are those that meet the needs of the customers and thereby provide product satisfaction. This corresponds to fitness for use.
2. Freedom from deficiencies (this applies to both product and process quality).

Fitness for use addresses human needs at different levels:

1. **Technological:** A product's technical features mostly specified by the customer, such as hardness or dimensions.
2. **Psychological:** Mostly subjective but appealing to the customers. In most cases, not very well defined but felt by the customer, such as finish or elegance.
3. **Time-oriented:** Mostly related to the performance of capital assets, such as availability, reliability, maintainability or throughput/manufacturability.
4. **Contractual:** Related to support and service provided during the usage of the product, such as warranty.
5. **Ethical:** While the psychological aspects pertain to products, the ethical aspects address the quality of service provided in the entire product life cycle, such as courtesy, honesty, integrity or responsiveness.

The most common perception of quality is at the product level. This has changed. Quality is less likely to be a differentiator and more likely to be considered a baseline requirement to be able to secure business. Customers generally treat basic product quality as a given and have evolved to equally value their overall experience across the complete product and service life cycle interactions.

Customer requirements, as per Noriaki Kano, can be classified into three groups:

1. **Assumed requirements/dissatisfiers/must-be attributes:** These are the features/requirements that are often taken for granted. Their presence is not even noticed,

but absence results in significant dissatisfaction. For example, if we go to a conference we expect a complete registration package and compliance to the agenda. Any missing piece/speaker no show/delay could cause significant dissatisfaction. We do not state that we are expecting the above, but we assume we will receive it.

2. **Stated requirements/satisfiers/one-dimensional attributes:** The customer specifically states these requirements. The more they are fulfilled, the more satisfied the customer would be. For example, when we go to buy a car we specify what we need in terms of the features, color, style and budget. If a dealership is able to identify a car fulfilling all of our stated requirements, we are extremely satisfied.

3. **Delighters/attractive attributes:** In the case of delighters, the customer is getting something that was not expected, and it provides a significant value. Since the customer is positively surprised, they are wowed. These wows often result in increased customer loyalty. For example, when we attend a conference and the book signed by the author is given away to all the attendees, this would delight the attendees. The challenge with delighters is that one needs to continually think about the next wow factor since the current wow factor(s) gets copied by the competition. Eventually, the differentiator evolves into an assumed future requirement.

Figure 1.3 depicts the Kano's model where the x-axis provides the extent of fulfillment of the requirements and y-axis provides the degree of satisfaction. Note that the "Must be" and "Delighters" are nonlinear in nature due to the emotions involved while stated requirements are linear in nature. This means the customers are extremely dissatisfied when the basic requirements are not met. At the same time, customers are extremely delighted when we are able to provide the delights. However, in the case of stated requirements, there are no positive or negative emotions. It is just the degree of compliance to those criteria. We shall be referencing Kano several times in this book.

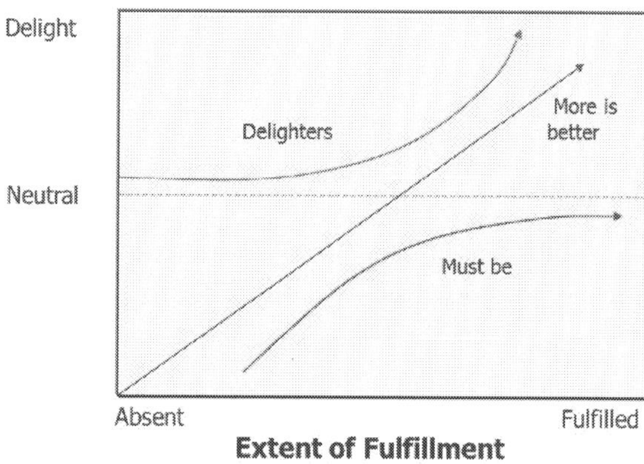

Figure 1.3

Businesses who fail to meet basic requirements now risk negative publicity with the advent of social media. Customers regularly "Yelp" their issues (real or perceived) to the world at large. Conversely, delighters can be a tipping point for success with the use of other forms of social media. A reality, though, is that a dissatisfied customer tells *"a lot of people"* about their experience while delighted customers tell *"only a few people"* about theirs. In fact, there is a book titled *Satisfied Customers tell Three Friends, Angry Customers Tell 3,000: Running a Business in Today's Consumer-Driven World* shown in **Figure 1.4**.

Figure 1.4

Some organizational strategies seem to focus heavily on one core "internal customer" group: the shareholders. This shows up pretty clearly in most "trifecta" strategic plans: Increase the revenue/ footprint, increase the margin and convert the cash. Sometimes there will be the incorporation of the needs of the external customer with traditional QCD (Quality, Cost and Delivery) objectives; however, these are often used as tactics to support the strategic trifecta.

There are several business evaluation systems, such as the Baldrige Performance Excellence (across most of North America and adopted by more than 90 countries), AME's Manufacturing Excellence Award and the newly updated ISO 9001:2015 standard, that incorporate a broader recognition of an organization's "stakeholders." From an ISO 9001:2015 perspective, these "interested parties" include the customers, suppliers, employees, the community and shareholders.

Customers can be defined as both internal and external. Internal customers can be interpreted within the vertical integration of the organization, including the employees and shareholders. For quality to be delivered externally it must be effective internally; or the processes will be inefficient resulting in less than optimal margins. The quality of systems that are perceived by our internal customers, our employees and our subsequent interactions with them, become a leading indicator of the quality of the product/service delivered to our external customers.

A ***great*** QLMS without certification is more impactful to the business than an average quality system with certification. An organization needs to have a bias towards a simple but highly effective (delivering results) and efficient (with minimal resources to maintain it) system. Teams need to focus on the reason why they are in business and develop a process map with clear identification of the internal customers across the value chain. As a business, our goal is to make money while providing exceptional products and services using effective and efficient systems.

Sophisticated customers care about the total cost of ownership. Since each of us is a customer, how would we like an organization to deal with us? If we look at all the aspects in relation to the product and service interaction, it generally starts with the service. Let us look at some key aspects from the life cycle perspective of both service and products.

Service Life Cycle

The service life cycle consists of non-product related customer experience touch points and interactions with the organization's brand. In most case, it starts with the product awareness through marketing or referral, product acquisition via sales, getting answers to the general questions during usage, getting problems resolved via service support and potentially product disposition in the event of a non-conforming situation. The efficiency and effectiveness of these processes are equally critical.

From an Accuride perspective, the functional teams took an encompassing approach: internal sales, external sales and marketing and engineering.

1. **Making Customers Aware (marketing process) and Acquiring Them (sales process):** The quality of marketing efforts and the promises made in customer acquisition are critical touch points for the customers to build a trusting relationship. While demonstrated QCD performance is key before some of these discussions can take place, customers are busy and need to build trust quickly. Many customers work with a Long Term Agreement (LTA) model, which can last from three to five years. Either you're in or you're out. It's a long and cold time in-between if you're not in. Prior to its Lean journey, Accuride lost some important LTAs. Now, with focused efforts on QCD and re-establishing trust, many are being renewed. Once customer agreements are in place, subsequent transactional processes for product acquisition need to be as effective (correct) and efficient (fast/easy) so that there is not a procurement burden.

2. **Answering the Questions (support process):** The quality and speed of answering customer questions during the Request For Quote (RFQ) process is another touch point. These sessions provide opportunities on how to develop additional value propositions that may be incorporated beyond just the delivery of product. For example, Accuride's Lean teams have partnered with several Original Equipment Manufacturer (OEM) customers to co-develop kanban systems in their receiving markets to minimize their on-hand inventory. While their on-hand inventory levels went down, in most cases by half, the sales volume remained the same, and the turns increased. In none of these cases did Accuride's inventory go up. Support teams need to anticipate the types of questions that may arise during the entire product life cycle to prepare for accurate and timely answers. There is a need to treat all the touch points

as an opportunity to build stronger partnerships.

3. **Handling the Problem (service process):** Customer problems are linked to a deficiency in the process or a misunderstanding of the product. Either way, the process failed to comply with the customer's anticipated needs; either the requirement was not anticipated, or there was an error in how the product was utilized. These situations provide opportunities to understand product, processes and train for issues that may arise and then develop countermeasures to manage them. After resolving the circumstance, these opportunities are to be analyzed for system improvement. **Figure 1.5** shows an example of how the internal sales team reviewed a set of customer satisfaction data and then tracked the effectiveness of the corrective actions that were applied (**Figure 1.6**).

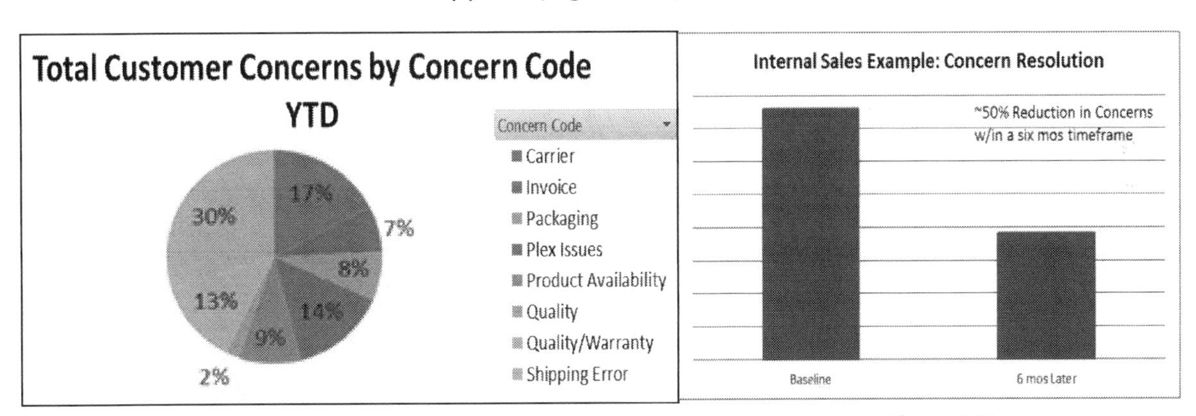

Figure 1.5 **Figure 1.6**

If the problem is satisfactorily resolved, we can build a stronger relationship with the customer. Sometimes customers can negatively take advantage of this and a balance needs to be struck. Team members handling these events should be empowered (within certain limits) to take whatever reasonable actions necessary to support the customer in order to maintain and improve the business relationship. These actions can potentially be a "Delighter" or wow factor to build customer loyalty and additional perceived value. Many customers leverage some type of "system" scorecard blending in QCD and other factors. This composite score reflects the overall customer experience and sometimes does not capture the additional intrinsic values that may be provided. Overall, while heavily expected, product quality in itself is only one component of the customer relationship.

4. **Product Disposition (service process):** The extent of product support during the entire product life cycle needs to be planned for and determined in advance. If it is not a part of the customer contract, a socially responsible organization may consider planning for how the customer will dispose of the product after its useful life is complete. Accuride's steel, aluminum and cast iron products are highly recyclable and customers usually sell them for

their metal content. Accuride's steel wheels, Steel Armor™ and EverSteel™, though, have had a significant coating technology applied to them that has enabled superior longevity and corrosion resistance. These technologies extend the life of a coated steel wheel for years beyond what is commonly seen in industry today. The stripping and refurbishment process (repainting) of steel wheels is very environmentally unfriendly and must be done when the original coating begins to wear. Accuride wheels with these new coating technologies enable customers to delay that refurbishment for years; including the sidelining of the truck and the labor to remove the tire and wheel assembly, remove the tire, paint the wheel, replace the tire, etc. Through such coating technologies, Accuride continues to focus on opportunities to provide a more positive customer experience.

Product Life Cycle

1. **Product Installation:** This aspect focuses on the quality contacts associated in activities, such as shipment, assembling components, product/process run-off, etc. In the automotive world, there is a process for this called Advanced Product Quality Planning (APQP). **Figure 1.7** shows a generic APQP process (Other companies have different variations of this theme).

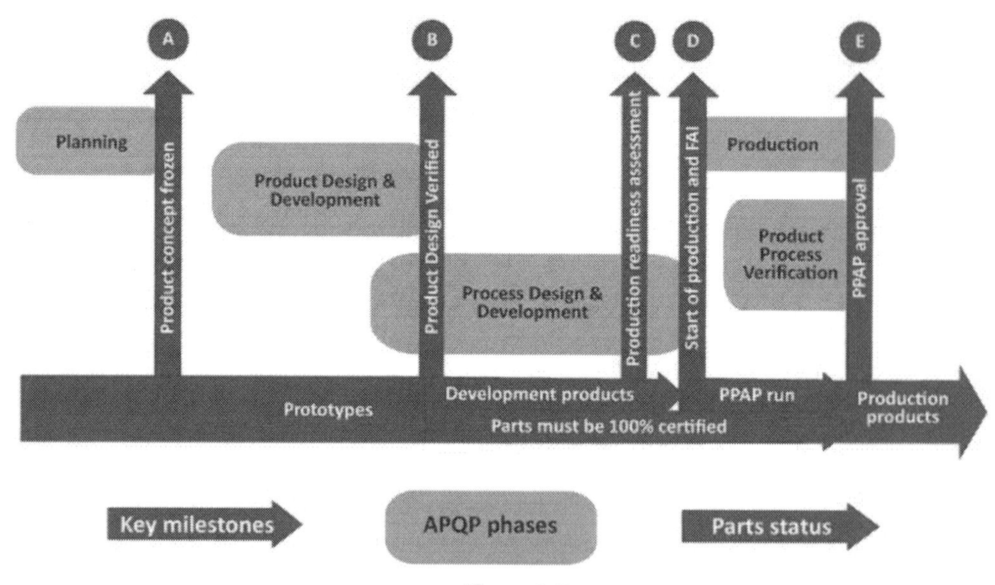

Figure 1.7

A subset of APQP is called Production Part Approval Process (PPAP). These processes are described and prescribed for OEM automotive customers via a system now called IATF 16949:2016 (The International Automotive Task Force (IATF) updated this global document from the Technical Specification (TS) 16949:2009). These process steps are defined in a series of "blue books" issued by the Automotive Industry Action Group (AIAG).

Accuride built upon the APQP foundation and developed an enhanced process called AccuLaunch™ shown in **Figure 1.8**. This stylized swimlane mapping format shows the various sub-processes across the five stage gates typical in an overall launch flow process.

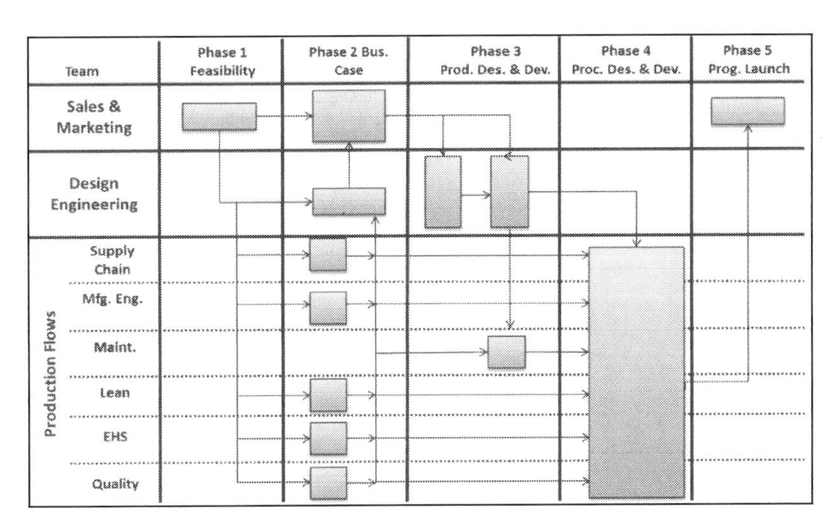

Figure 1.8

AccuLaunch™ incorporates additional Lean, EHS, Material Handling and Maintenance needs in the overall APQP process. We call it APQP on steroids. This cross-cutting process flow is captured via a "swimlane" T-VSM, which enables the team to see how the various functions interact, deliver and receive information across the product and process development cycle (See Chapter 4). There are QCD features of this process as well:

Quality: From a run-off perspective, how many of the critical characteristics meet the required initial process capability (Cpk)?

Cost: Did the finished product/process meet the costing estimate? Was it over or under the costing plan, enabling the business and the customer to meet their financial objectives? This is expressed as a percentage.

Delivery: Across the "stage gates" of the process (there are five identified in APQP and AccuLaunch™), did it advance on time and deliver the initial results as planned?

Product installation may include service opportunities, such as training on how to use the product as intended. These opportunities may provide additional intrinsic value to the customer. If necessary, how will this training be deployed? Face to face? Internet/YouTube video? Pamphlet inserts? Company personnel (technical, customer service, sales) may need training to affect knowledge transfer. For new products, internal training is a part of AccuLaunch™ to ensure effective customer communication. For example, some customers regularly refurbish their steel wheels in a given timeframe, necessary or not. With training, some of these customers are now experiencing the benefit of delaying the expense of refurbishment if it is not warranted.

2. **Product Appearance/Attractiveness:** Customers expect both high value/performance and appearance attributes to meet their needs. For example, we would not accept a car with a scratch on the door; it may work fine and be safe, but it does not meet the attribute criteria. At the customer scorecard level, regardless of the severity, all Defective Parts Per Million (DPPMs) are created equal. Appearance features need to be agreed upon and understood in advance of delivery. This evaluation is a part of PPAP. A formal evaluation, called an Attribute Agreement Analysis (AAA), can be conducted to "calibrate the human gauges" performing the inspection. This can be a complex process with several iterations because sometimes the customer is not able to articulate the appearance parameters that are needed. Most of these elements tend to fall under assumed/basic requirements of the Kano's model.

3. **Product Performance/Cost of Operation:** Whether referring to capital equipment or components, customers need to manage their handling, possession and application of the product in *their* planned manner. This can evolve the product integration beyond that of traditional form, fit and function of the unit at hand. This can include non-deteriorating storage and handling, returnable dunnage/packaging management, ease and manner of product installation, matching to custom built tools/jigs/fixtures, ability to meet build cycle times, ease of serviceability when in the field, retention of product traceability across the supply chain, etc. These aspects need to be reviewed in advance and are typically evaluated as a part of the APQP process.

   A common Lean KPI for evaluating cost of operations is OEE (Operational Equipment Effectiveness). OEE is a leading indicator to margin that evaluates whether or not we are utilizing our assets in such a way that we are getting quality products close to an ideal production rate from the asset during the planned run time. OEE is the product of the following key elements:

   OEE = Availability % x Throughput % x First Pass Yield %

   a. **Availability/Uptime:** This addresses the asset *availability loss* and is calculated as a percent of the time the assets were available as compared to the time they were scheduled to run. The formula for availability is:

      Availability = (Scheduled Run Time – Downtime)/Scheduled Run time

      Downtime is limited to unplanned downtime because planned downtime is factored into the scheduled run time. The facility leadership needs to determine the planned downtime for their assets for optimal performance as this will affect the total available time for the asset to run. Scheduled breaks are taken out of the time to calculate the scheduled run time.

Downtime is normally composed of two key factors that need to be analyzed separately: maintenance related (unplanned breakdowns, unavailability of spare parts) or non-maintenance related, such as changeover time, waiting for components, unavailable employees, unplanned meetings, etc.

b. **Throughput/Efficiency:** Throughput defines the *speed loss* of the asset. This can be a controversial element in OEE. If not addressed correctly, it can give a false sense of accomplishment:

Throughput = (Number of pieces actually produced in the available time) / (Number of pieces that should have been produced in the available time if the assets ran at the ***ideal*** *cycle time)*

The challenge lies in the word "ideal" and how honestly it is defined. We have seen organizations reporting more than 100 percent efficiency. This is not possible as the asset can never run faster than the ideal cycle time. If the asset has been modified, then the ideal also needs to be modified. The idea is not to make this number look good but to identify opportunities for improvement. Another common situation where Throughput exceeds 100 percent is when averages are used across part families to accommodate for differences in the ideal cycle time. The gap occurs when the production mix does not match the "average." This averaged ideal is used for convenience. When an average is needed, a cap of 100 percent should be reported when a mix in production causes an excess of 100 percent.

c. **Quality/First Pass Yield:** This element addresses product *quality loss* and is calculated as a percent of the good units made compared to total units made during the available time and speed. Only the good products can be recognized. Scrap and/or *Rework* are removed from this count (Many try to forego the "rework" counts, as they believe that only the ending scrap should be counted. This is a classic cheat and thwarts the true visibility of the losses being incurred in the system. Someone has to move, handle and manage those units; margin loss is going to be incurred). The formula for the First Pass yield is:

First Pass Yield = (Number of ***good*** *pieces produced at actual production speed when the assets were running)/ (Total number of pieces that were* ***produced*** *at actual production speed when the assets were running)*

In some batch processes, the quality assessment of the goods can be delayed after a secondary processing. For example, in Accuride's foundry operations, the quality portion of the OEE for mold lines is calculated after a finishing operation. Closed molds (molds that were not poured) are assessed as a yield loss.

4. **Product Disposition:** As a product reaches its end of life, an evaluation of who needs to disposition the product should be considered with respect to social responsibility and planned economic benefits to the key stakeholders. Accuride's products are wholly recyclable and efforts are made to ensure that coatings and supplementary products are as enhanced as possible with an eye towards sustainability.

Cost Of Poor Execution (COPE)

Most people are aware of Cost Of Quality (COQ) or Cost Of Poor Quality (COPQ). Juran developed a COQ model shown in **Figure 1.9** with four types of quality costs:

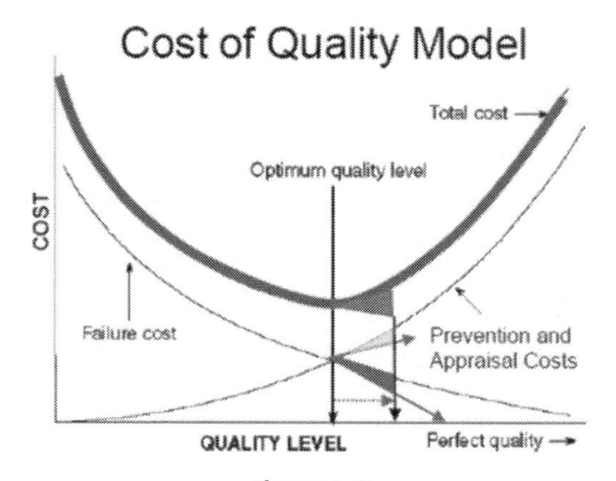

Figure 1.9

1. **Prevention costs:** Costs associated with preventing the poor quality.
2. **Appraisal costs:** Costs associated with ensuring that only the good product is produced.
3. **Internal Failure costs:** Costs associated with suspect product that is identified and contained within the four walls of the organization.
4. **External Failure costs:** Costs associated when suspect product has been shipped to the customer.

While all four types of costs above make the COQ, the last three make the COPQ. There is a sample listing in **Figure 1.10**. Lean impacts the flow of product and information resulting in the reduction of expenses in each of the four COQ categories. Accuride uses a KPI called COPE. While COPE incorporates many of the COPQ losses, it also monitors ***both*** *product and process losses*. COPE includes losses due to Muda (system waste), Mura (fluctuation and variation in the resource demand) and Muri (burden on people and equipment). **Figure 1.11** shows Accuride's overall COPE trend and results with respect to what we've had to "cope with" to recover margin. COPE is measured as a percentage of the Cost Of Goods Sold (COGS). It is not compared to revenue for the following reasons:

1. If measured as a percentage of revenue, it gets diluted where the trends generally do not reflect actual performance due to sales mix and seasonality.

2. A COGS denominator makes the margin impact visible, relevant and recoverable. Facilities may have differing loss leaders (or margin eroders). Using the Pareto principle, Accuride sites are empowered to evaluate their top four to five loss leaders, which cover 75-80% of the overall COPE. Typical COPE contributors are:

   1. Scrap
   2. Overtime (unplanned labor)
   3. Shrink (usage variance)
   4. Unplanned maintenance labor
   5. Rework
   6. Excess and obsolescence
   7. Lost capacity
   8. Premium freight and so on.

External Failure
Customer complaint investigation
Returned goods (warranty, shipping, labor to handle goods)
Recall costs
Liability costs
Penalties
Premium freight
Pricing errors
Internal Failure
Scrap
Inventory losses
Rework, Repair
Waiting
Premium freight (inbound)
Material losses (MUV)
Disposition losses (labor, delivery,...)
Corrective action/investigation processes
Re-inspection/Re-test costs
Over-production (sub-assembly)
Material Review Board (MRB)
Design corrective action
Appraisal Costs
Document checking
Purchasing appraisal costs
Receiving/Incoming

Appraisal Costs cont.
Qualifying supplier product
Product/Process inspection
Financial audits
Inspection and test materials
Measurement equipment expense
Maintenance and calibration labor
Outside certifications
Laboratory testing/support
Prevention Costs
Marketing research
Customer surveys
Contract/Document review
Design quality review/testing (PPAP)
Field trails
Supplier quality reviews
Operator training
SPC
Quality Admin salaries/expenses
Quality performance reporting
Quality system audits

90% of the losses are "under the water" and intrinsically hard to capture!

Source: Google Images

Figure 1.10

DFSS, Lean and Six Sigma are key tools in reducing COPE. So is an aligned functional team. COPE transcends all functional teams in an organization. As the various components are understood, their processes can be reviewed for improvement. One key leading indicator for scrap that Accuride uses is called Key Process Indicator Variables (KPIVs). This is a master list across the organization that identifies every critical characteristic for its Cpk (See Appendix A). If a Cpk is greater than 1.33, then the process is expected to yield good results. If it is

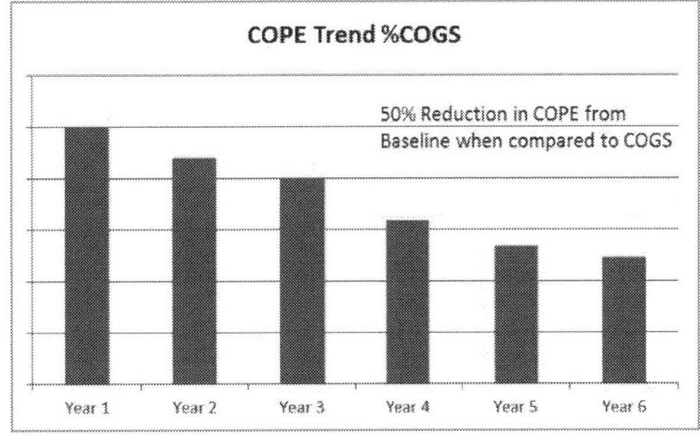

Figure 1.11

less than 1.33, then containment measures are needed to mitigate erosion of effectiveness, efficiency and margin. The %KPIV monitors the percentage of all variables that meet a Cpk of 1.33 or greater. Features with a Cpk < 1.33 are contained with aggressive corrective actions and countermeasures. Some %KPIV evaluation approaches are listed here:

1. Ensure that the data being collected is correct: A Measurement System Analysis (MSA) may be needed. Remember that there are effective analysis methods for both variable and attribute characteristics. The Gauge Repeatability and Reproducibility (GRR) method is for variables gauging and the AAA is for attribute features.
2. Look for an applications error in the statistical approach: Sometimes single sided specifications are treated as double sided, causing an improper Cpk result. For example, flatness typically has only one Upper Specification Limit (USL), with the intent to be as close to zero as possible. If the process average is close to zero, and the process is treated as a normal distribution, then there will be an error in the results.
3. Check for Normality of the data by making the histogram: If the data is not normal, different methods may be needed to properly evaluate the Cpk and process risk.
4. Cpk is close to 1.33: Review the process for the potential of shifting it in a desired direction.

The more stable the processes are, the less we have to "COPE" with. For an organization with multiple manufacturing and distribution sites, the opportunities in the COPE elements will be different from site to site and different actions will be needed. When something is measured, actions are required to drive the change or improvement or the effort is wasted. Dr. James H. Harrington, shares this sentiment in **Figure 1.12.**

Measure

Don't MEASURE it if you aren't going to RECORD it
Don't RECORD it if you're not going to GRAPH it
Don't GRAPH it if you're not going to ANALYZE it
Don't ANALYZE it if you're not going to take ACTION

Why Measure it if you are not going to take action?

Figure 1.12

Key Takeaways

1. A 50% average reduction in lead time can result in a 50% increase in productivity and an average reduction in controllable cost per unit by 20%.
2. Lead time reduction is accomplished by reducing the process variation (DFSS and Six Sigma) and eliminating the waste (Lean). Both elements are necessary.
3. Leaders communicate to inspire actions and support their teams to accomplish goals.
4. Leadership is required at all the levels as change moves at the speed of trust.
5. Quality is defined by the customer. Use all customer touch points to strengthen the relationship.
6. Reduce the COPE by identifying top product and process losses.

Chapter Two: Strategic Planning

Several years ago, a USA Today "Snap Shot" stated that only 3 percent of North American businesses shared their strategies with their personnel.

It is very unfortunate that all of the hard work putting those strategies together does not get translated into effective actions simply because the individuals who need to make them happen are not made aware. It is therefore critical that business leadership shares as much as possible with their associates so that, at the very least, they can be in a position to help the company to get to wherever "there" is. Personnel with clear goals and actions connected to the overall business generally have higher levels of engagement. Higher engagement leads to happier and more productive people. Per Shawn Achor's *Happiness Advantage,* businesses with happy people have 37 percent more sales, are 31 percent more productive and are 10 times more engaged!

So, why is C-Suite worried about sharing key strategies with its personnel? There are generally two reasons:

1. Personnel may not be in a position to really understand what the strategies are and why they are needed.

2. The competition might find out.

As for the first, it is up to the management team to create that translation of the strategic whys. In *How the Mighty Fall*, Jim Collins outlines how businesses decline and identifies many signs for businesses to recognize as to whether or not they are on this downward spiral (**Figure 2.1**).

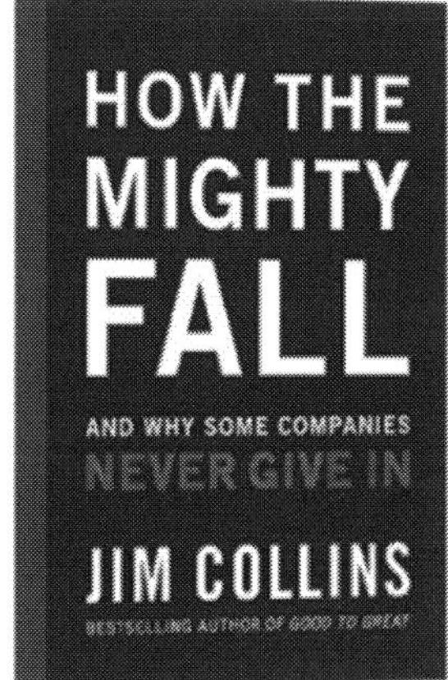

Figure 2.1

By not sharing the rationale behind the business strategy, the leadership team runs this risk as well as the possibility of perpetuating confusion: even among themselves. As for the second, it is highly likely that the competition already knows what the next logical plays are. Of course, the amount of detail that is shared may need to be filtered. Nevertheless, the main objectives need to be shared.

If we think of some of the most successful Lean businesses today, Toyota, Autoliv and OC Tanner often come to the forefront. All of them welcome people to come and see their methods and processes. Their strategies are widely broadcast from their leadership and disseminated across the workforce. Most use some form of Hoshin Kanri to cross-connect the key business strategies to the functional initiatives. Bill Waddell stated it well in his blog on *Toyota's Eyes*, "Seeing your business through the same lens won't lead to the changes you hope to make. Instead, you must learn to look at the business through Toyota Eyes..." (Or, through the eyes of whomever you are looking to benchmark). Toyota is generally referenced as synonymous with Lean since several modern Lean concepts have their origin in the Toyota Production System (TPS).

We need to think of the following questions:

1. How do we force ourselves to look at our organization through that different lens?
2. How do we then apply that feedback to the business strategy?
3. What are the key process steps that are needed to effectively link our functional initiatives with that of the business strategies?
4. What are the major steps needed to harness that aligned power to develop a Lean organization?

Management professionals of all business functions and at all levels of the organization need a high level plan of approach on how to:

1. Objectively look at the status of one's own business from a Lean perspective.
2. Develop a Hoshin Kanri linking the business' key strategies to that of the organization's functions.
3. Select and identify KPIs of success.
4. Leverage the Hoshin Kanri to execute Lean across the business enterprise through four systemic phases.

By managing through these systems, a business professional can more objectively develop their own process that will better enable their organization's success.

Benchmarking... Looking Through the Other Lens

How does one look at their business objectively to see it for its possibilities? There are generally two paths business teams can consider. They can:

1. Pay for a third party evaluation and secure support from a consultant organization.
2. Find a way to conduct their own benchmarking and Do It Yourself (DIY).

Both have strong pros and cons. To pay for a third party evaluation can be quite costly. But, it may have a shorter overall lead time than the DIY method. There is the risk of selecting a good partner that will not only help to identify the opportunities but also help the team to get kick-started on its journey without becoming a permanent cost fixture in the organization. There are many reputable and strong third party Lean practitioner resources out there. They seem to follow a general approach:

- They will aid in the initial assessment of the organization and facilitate the creation of the initial current state Value Stream Maps (VSMs).
- They will provide the business with basic Lean and systems training.
- They will stand back and let the team members of the business do the legwork, execute the physical actions, while the third party monitors from the sidelines. They are, after all, teaching the business "how to fish." If the third party is executing the tasks and the implementation of the systems, the business' ownership of the improved processes will not generally hold up over time.
- As the business evolves through its phases of improvement, the third party will bow out, leaving the organization to stand on its own. Depending upon the organization's size and complexity, this partnership can last from two to four or more years.
- Most critically, the third party may fire the organization, its own customer, if it is not ready and/or able to effectively commit to the journey. How many of our organizations would willingly turn away business?

Third party partners can be powerful and effective when the leadership team feels it needs this level of support (Accuride did this at the start of its journey and still leverages them for a "fresh eyes" review).

The DIY method also has its own pros and cons. Depending on one's perspective, a main detractor may be the longer lead time for implementation. And time, as we know, is money. Regardless, as long as the organization starts on its journey, it can get there with the commitment of the senior leadership team. One strong positive, though, is that the organization's unique culture is cultivated along the way. People who are not comfortable engaging in the process often self-select by leaving. Aside from not having to provide severance packages, this approach can be smoother for a company.

There are several large and inclusive Not-for Profit Organizations (NPOs) that can help the DIYer. These NPOs often host a variety of conferences, webinars and site tours. They provide local training and mentoring support with benchmarking opportunities. The level of engagement with these entities is up to the organization. You can do as much or as little as you want. These organizations have survey tools that businesses can conduct internally. As companies work on their Lean journey, they can compete with others via these entities' recognition programs. When an organization is ready to be objectively evaluated, they can apply for recognition. And, essentially for the cost of Travel and Expense (T&E), these organizations will send Subject Matter Experts (SMEs) to the business and evaluate its level of Lean execution. Just a few examples include:

- **The Baldrige Foundation**. The Baldrige Foundation manages the Baldrige National Performance Excellence Program in the United States. The President of the United States presents this award. The Alliance for Performance Excellence manages the Baldrige Performance Excellence Award at the state level. Almost every state participates in this program. This process is an amazing way to learn about how to develop, implement and manage performance excellence systems and then participate in site assessments to see how others apply the techniques. The American Society for Quality (ASQ) supports the Baldrige Performance Excellence process. Accuride has personnel that participate in this program.

- **The Association for Manufacturing Excellence (AME).** AME's recognition program is the AME Manufacturing Excellence Award. Shown in **Figure 2.2** are the press release titles showing Accuride's four sites that have won this prestigious award. AME focuses on developing a mentoring relationship. Like Baldrige, their rigorous application process forces an organization to be objective and introspective in its review of itself. Having gone through the process four times, each site, as well as the overall business, has made significant improvements in performance. The subsequent SME feedback was also highly valued. Accuride is a corporate member of AME.

2016 AME Excellence Award Recipients

2015 AME Manufacturing Excellence Award Recipients

PAST RECIPIENTS

2014 AME Manufacturing Excellence Award Recipients

Figure 2.2

- **The Shingo Institute**. Shingo's recognition is the Shingo Prize for Operational Excellence. Shingo is globally regarded for its in-depth evaluation of an organization's continuous improvement system.

The Shingo Prize is the world's highest standard for operational excellence.

- The **Japanese Union of Scientists and Engineers (JUSE)**. JUSE manages the Deming Prize, which is a global quality award that recognizes both individuals for their contributions to the field of Total Quality Management (TQM) and businesses that have successfully implemented TQM. It is the oldest and most widely recognized quality award in the world and was established in 1951.

JUSE
Union of Japanese Scientists and Engineers

- There are several Lean system self-assessment tools (including those provided by the sites above, Bill Waddell's 100 Question Lean Survey (see bibliography) is very intuitive.

Regardless of the method of support that is chosen for a business' Lean journey (third party or DIY), either should enable an organization to rapidly better itself so that it can provide more value to its customers and become more competitive.

Hoshin Kanri... Linking Business Strategy to Functional Initiatives: A QLMS Example

Now that the business is on its way to becoming more agile via Lean, the need for the functional teams to link their initiatives to the business strategies come into play.

It is key that the functional professionals proactively align themselves to the organization's strategies. This demonstrates the relevance of their systems to the business at hand. While all functions need to be aligned, for simplicity, this section will specifically reference the QLMS of the organization and their connection to *sample* business strategies. For this partnered evolution to happen, the QLMS needs to have several areas of focus:

1. Sustain and build upon effective QLMSs to meet product *and* process requirements.
2. Recognize and act upon the introspective responsibility to reduce system costs in order to maximize margin potential.
3. Develop and leverage effective methods of flow that enable product and process improvements to profitably grow the business.

Many organizations only focus on a subset of actions under item number one: they place their emphasis on "Lean manufacturing" in the operations environment and the "quality of the product." While this is truly necessary, this version of the QLMS is only looking at a narrow slice of the business' total opportunity. By incorporating the transactional support processes and items two and three into their repertoire, quality and Lean professionals can link their objectives to that of the organization's strategy. Daniel Zrymiak's article "Understanding Governance Within Organizational Excellence and Management Systems" outlines some potential approaches. The QLMS structure needs to be aligned to the strategic plan within the governance structure.

One powerful methodology is Hoshin Kanri. This step-by-step planning process often uses an iterative matrix model to connect the organization strategies to the initiatives of supporting functions. Those initiatives are broken out into tasks and key metrics are defined. As the matrix is developed, an interactive process of learning and alignment occurs across the organization's leadership team. This "catchball" process (a back and forth byplay) helps to align the organizational functions on the development and execution of their functional strategies. Hoshin Kanri has evolved into many variants.

Grace Duffy's article, "How Group Decision Making Helps With Functional QMS Strategic Planning," provides further insight into how to develop the general content of the strategic plan. Hoshin Kanri is a robust process; described herein is a *simplified concept* as to how the strategies, as developed by an organization's finance and governance team, can be linked to the initiatives of a QLMS. Note: It is assumed that the front-end activities of conducting the environmental scans completing the overall strategic analysis and the development of the key strategies have occurred. **Figure 2.3** shows a simplified model of the center of a Hoshin Kanri. In this example, there are four sections.

- The six o'clock (bottom) position lists the organization's strategies (Assuming that they have already been developed). Following both Performance Excellence formats (Baldrige) and the newly upgraded ISO 9001:2015, full strategies are required to address various "customer" groups or "interested parties" including: **S**upply Chain, **E**mployees, **C**ustomers, the **C**ommunity and **S**tockholders (SECCS). The challenge for the quality and Lean professional is to align the QLMS processes and results to that of the business; whether or NOT Lean systems are specifically mentioned in the resultant strategies. **Figure 2.4** shows an example of some potential business strategic priorities.

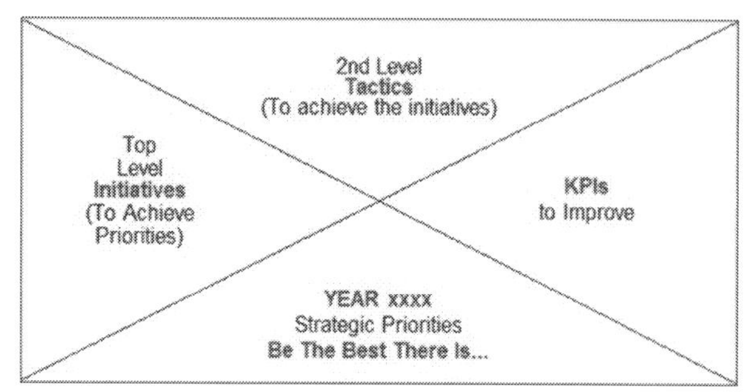

Figure 2.3

- The nine o'clock (left) position lists the relevant QLMS initiatives. As each initiative is identified, determine how each one matches with the various strategies. Correlation dots help to show the alignment of the initiatives to the strategies. This minimizes the number of initiatives the team needs to have in order to be in alignment with the objectives. In the example, if "scrap/defect abatement" occurs, then strategies one, two, three and five could be positively impacted. ***Scrap abatement is key for Lean systems... Process flow cannot occur when defects are occurring*** (Whether those defects are encapsulated in product or in transactional data). When communicated to the finance and governance functions, this linkage reinforces the value proposition that the QMS is bringing to the team. **Figure 2.5** shows this linkage of strategic priorities with initiatives.

1. **Stakeholder**: Grow Profitable Sales by X% by Y Date
2. **Stakeholder**. Grow Margin by X% by Y Date
3. **Customer**. World Class Quality, Delivery
4. **Employee**. Grow Satisfaction by X%, Minimal Unplanned Turnover to Y%
5. **Supply Chain**. World Class Quality. Delivery
6. **Community**. Grow Social Responsibility, Meet/Exceed Compliance Requirements

Figure 2.4

- The twelve o'clock position lists key tactics that will be executed to accomplish the initiatives. Correlation dots are again used to show which tactics support which initiatives. This method identifies tactics that are impacting multiple initiatives. Again, this alignment drills down to the original strategies. It minimizes the number of tactics that are finally developed. In the example shown in **Figure 2.6**, the Tactics for the year are aligned to the Initiatives. A1 demonstrates how a tactic can span several initiatives. This cross-cutting tactic enables the optimization of functional team resources and reduces "silos." A2 is also cross-cutting and impacts four initiatives, which in turn affects <u>all</u> of the strategies.

E. Compliance Systems Management	D. QLMS Systems Optimization: Transactional Processes	C. Lean/Continuous Improvement Integration	B. Customer Satisfaction	A. Scrap/Defect Abatement; Process Control	Top Level Initiatives (To Achieve Priorities) / 2nd Level Tactics (To achieve the initiatives) / KPIs to Improve / YEAR xxxx Strategic Priorities Be The Best There Is...
	●		●	●	1. Stakeholder: Grow Profitable Sales by X% by Y Date
	●	●	●	●	2. Stakeholder. Grow Margin by X% by Y Date
●	●				3. Customer. World Class Quality, Delivery
●	●				4. Employee. Grow Satisfaction by X% Minimal Unplanned Turnover to Y%
●	●		●	●	5. Supply Chain. World Class Quality, Delivery
	●	●		●	6. Community. Grow Social Responsibility, Meet/Exceed Compliance Requirements

Figure 2.5

- The three o'clock (right) position lists KPIs. There are several ways to develop KPIs. Gary Cokins' article, "Fixing a Kite With a Broken String – The Balanced Scorecard," provides excellent guidance.

E. Compliance Systems Management	D. QLMS Systems Optimization: Transactional Processes	C. Lean/Continuous Improvement Integration	B. Customer Satisfaction	A. Scrap/Defect Abatement; Process Control	Top Level Initiatives (To Achieve Priorities) / 2nd Level Tactics (To achieve the initiatives) / KPIs to Improve / YEAR xxxx Strategic Priorities Be The Best There Is...
			●		B1. Conduct Customer Satisfaction Survey. Evaluate Results and Implement Corrective Actions based on Feedback
●	●		●	●	A2. Execute SPC Review of all Key Characteristics. Verify for Effectiveness
		●	●	●	A1. Define Black Belt projects, Implement and Sustain
	●		●	●	1. Stakeholder: Grow Profitable Sales by X% by Y Date
	●	●	●	●	2. Stakeholder. Grow Margin by X% by Y Date
●	●				3. Customer. World Class Quality, Delivery
●	●				4. Employee. Grow Satisfaction by X%. Minimal Unplanned Turnover to Y%
●	●		●	●	5. Supply Chain. World Class Quality Delivery
	●	●		●	6. Community. Grow Social Responsibility, Meet/Exceed Compliance Requirements

Figure 2.6

Benjamin Franklin said, "Watch your pennies, and the dollars will take care of themselves." This is wise advice when applying it to managing KPIs and differentiating between leading and lagging indicators. Essentially, if you only manage the lagging indicators, the system of management is largely reactive vs. proactive or predictive. For example, margin is a typical lagging indicator. It is absolutely a required metric from "the Street's" perspective. However, by managing its key inputs, such as scrap, rework, overtime, etc. up front, the team has a greater chance of yielding a desired result. Manage the leads, and the lags will take care of themselves.

Over time, there is some synchronicity with a 2-to-1 ratio of leading indicators to lagging. By having the team focus their efforts on managing those two leading indicators (the pennies), the lagging values (the dollars), usually manage themselves.

- In this example, both leading and lagging indicators are demonstrated. Linkage between the KPIs and the tactics can easily be seen in **Figure 2.7**. A powerful benefit to this visibility is that the impact of each tactic on a KPI can be understood and possibly predicted, enabling the organization's governance function to more clearly identify how the QLMS functions are providing tangible value to the team.

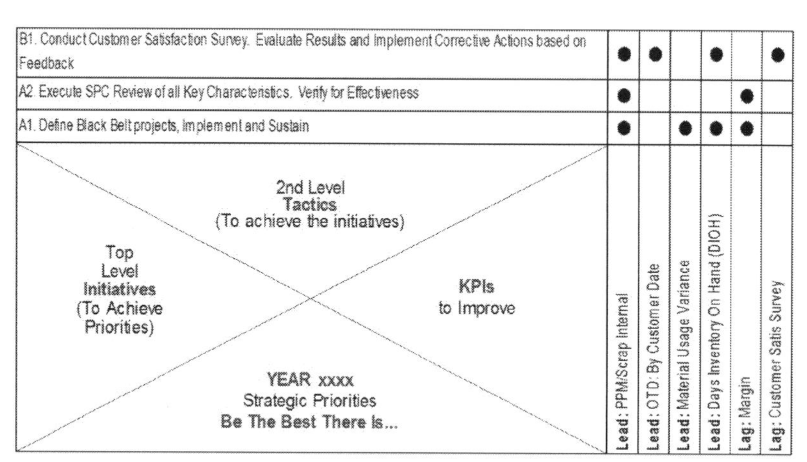

Figure 2.7

- When the full Hoshin Kanri is developed, the leadership has constructed a clear linkage from the strategy of the organization through the development of aligned initiatives and effective tactics. A highly simplified model is shown in **Figure 2.8**.

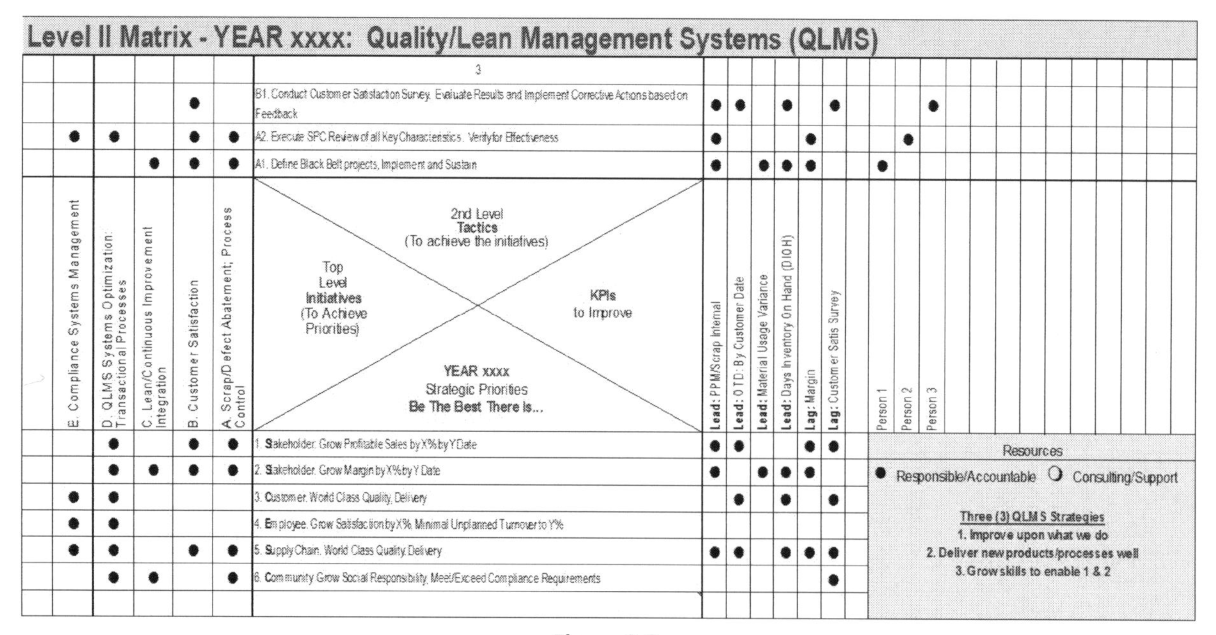

Figure 2.8

Once the Initiatives, Tactics and KPIs are defined, the process of *aligned execution* begins. While sustaining and/or building upon the compliance portions of the QLMS, the Lean and Quality Professional needs to continue to demonstrate the value they can bring to the organization. **This sample is inclusive of just two typical business functions.** A similar alignment should be made with the other functions of the organization, such as Supply Chain, Engineering, Human Resources, IT, Legal, Finance, etc. These matrices are referred to as Level II Kanris. Depending on the size of the organization, the Level IIs may be blended into one large document, or they may be retained independently by the functional groups as the tactics are executed. It is, however, important for the functional teams to see others' Kanris. This avoids a duplication of effort. The teams can collaborate on common themes to further refine the initiatives.

Key Process Indicators (KPIs)

The right hand side of the Hoshin Kanri identifies the KPIs of the organization. There are varying approaches on how to develop these. As mentioned earlier, a blend of leading and lagging indicators helps to control the front end so that the back end follows. Whatever the KPIs of the business are, it is quite helpful to monitor them in a graphical format to look for shifts, trends and cycles. This helps when determining the next course of action. Forrest Breyfogle's book *Business Deployment Vol. II: A Leaders' Guide for Going Beyond Lean Six Sigma and the Balance Scorecard* provides a detailed approach on how to break down KPIs and manage them in a predictive manner.

From a DIY perspective, the format in **Figure 2.9** is found to be most effective and inclusive when managing KPIs. This type of visual data is also referred to in several of the advanced QLMS models, such as Baldrige and AME. Some element details are:

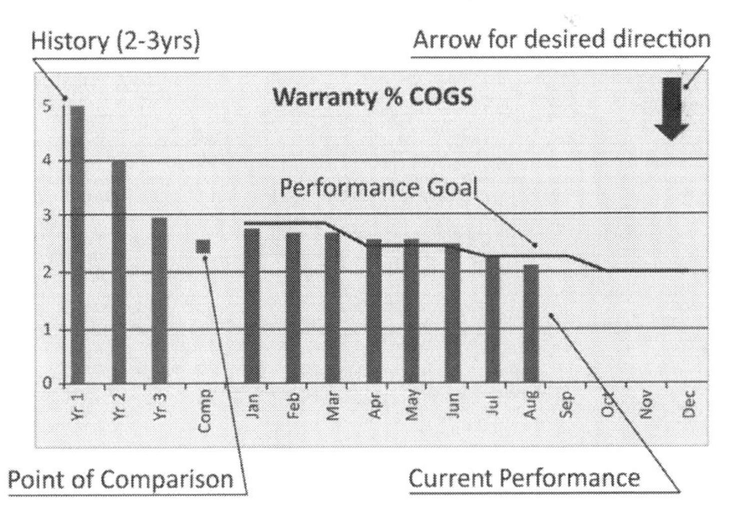

Figure 2.9

- Some form of **History** (if available) should be provided to tell the story of overall performance. It is a general practice to not go beyond three years because systems and metric definitions change over time. When applying for general performance awards, this history is often reviewed as part of the evidence of sustainability.

- A **Point of Comparison** (POC) should be provided where practical. A POC can be a benchmark reference but should not be confused with one if it is not truly the best in class for that measured feature. A POC allows for flexibility for comparisons to be made to any of the following sources or "levels" of data comparisons: International, National, State, City, Industry, Competitor, Inter-Company, Inter-Department, etc. The overall intent is to develop an objective comparison with another entity to see where the organization is really performing. POCs can work in two ways:

    1. If the team is performing negatively to the comparison, it can provide motivation and clarity that the comparative level can be achieved; if someone else has done it, so can they. In the non-Accuride example provided here, the team felt great when they went from five to four in their Warranty performance as a %COGS. A 20 percent Year Over Year (YOY) reduction! Without an effective comparison, a team may feel they've really hit the new low. However, when faced with the reality of a comparison, they are able to rapidly adjust their filter and look at the problem differently. From year two to three, they reduced it again by 25 percent, and yet, it still wasn't enough. They had sufficient motivation to know that there was a way to do this in order to recover their competitive position.

       The real challenge, though, was that in this example, the customer advised the business of a competitor's result in Year Two. The customer took a mandatory pricing reduction citing that the pricing was inflated due to the costs of the warranty. The business' choice was to forego the revenue or address their performance. They chose the latter. When they ended up exceeding the comparator and showed the customer, they were able to recover some of the pricing. One can only assume that the customer possibly shared the business' new performance level with the competitor to extract further cost downs.

    2. If, however, the team is performing positively relative to the comparison, it can provide a sales or marketing advantage that can be leveraged.

- A **Performance Target** or objective needs to be identified. This shows the team how the KPI is performing to the plan. The target can be flat or stepped as progressive goals are achieved.

- The **Current Performance** of the KPI is tracked on either a rolling-12 (last 12 consecutive months) or in a year to date format. The idea is to observe the overall behavior of the process over time.

- The **Arrow of Desired Direction** helps the reviewing audience to understand the performance intent. Which way is desired? Up, down or flat?

There are a few challenges in interpreting graphs at the *leadership* level:

1. Data is usually collected monthly; there are not usually "hundreds" of data points at "high check" frequencies to assess for performance. This is another benefit of using leading indicators.
2. People tend to "over-characterize" the performance of a process. Two points over average may be misidentified as a "trend." One point "higher than the rest" may be labeled as a "spike."
3. Most critically, many of these KPIs are NOT typically expected to be normally distributed. Depending on the type of KPI, it is often hoped that they will tactically trend (or shift) up or down. Margin, Revenue or Market Share perpetually going up may be considered to be a great thing. A flat-lined PPM at zero would also be quite nice.

By applying Shewhart's basic principles of control, (See AT&T's Statistical Quality Control Handbook), a less reactionary interpretation of the KPI performance can be made and enable people to focus more on reducing the underlying causes of variation. Some points on control:

- Per Shewhart, a shift has occurred when there are "eight points in a row above or below the average." Other references will cite five, six or seven points. However, the bottom line is that "two or three" points in a row above or below the average does not make a trend. Think of it this way, for a normal process, you should have an equal chance of getting a value above or below the average. The same odds as flipping a coin. If you flipped a coin twice, there's a good chance you could get two heads or two tails consecutively. The odds of getting the same face decreases dramatically with each consecutive flip. By flip five, six, seven or eight, the *possibility* is there that you'd still be getting a consecutive head or tail, but the *probability* of getting one is pretty close to nil. It's just not normal. At that point, you're looking pretty closely at that coin. Therefore, from a process change interpretation perspective, there really isn't any shifting going on unless there are more than five to eight consecutive points in a row above or below that average; depending on whose book you

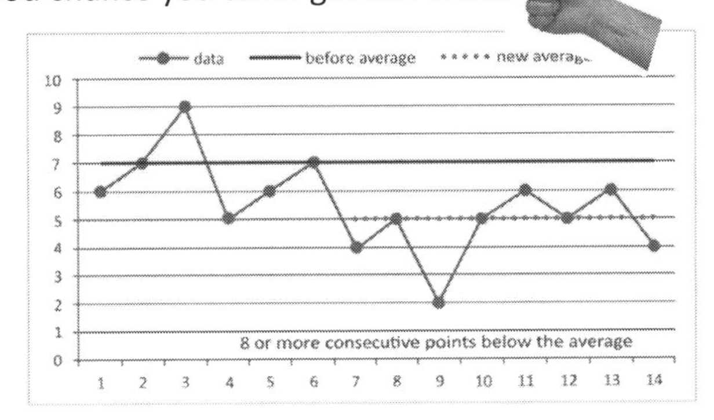

Figure 2.10

reference (**Figure 2.10**). You should determine for yourself what that five to eight "shift" trigger should be.

- A similar point is made regarding trend behavior. Two or three points going up or down does not a trend make. There needs to be six or more data points going consecutively up or down for the process to be considered trending (**Figure 2.11**).

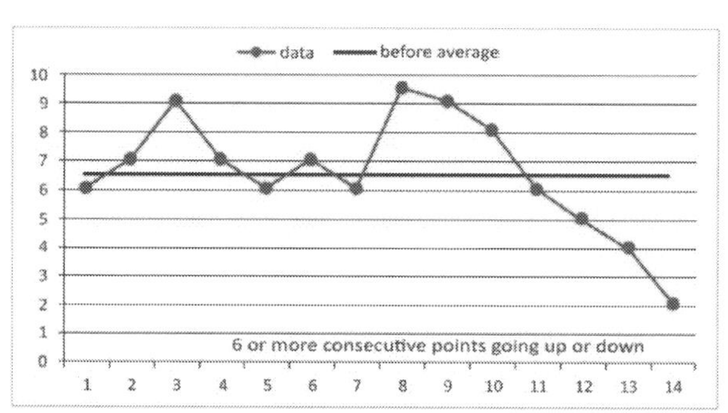

Figure 2.11

- A "spike" in performance can be considered for removal from the data set if there is a verified special set of conditions that caused the metric to behave that way. If a special cause is not identified, then that super high or low point may just be part of the normal variation of the process.

When talking about KPIs, businesses can sometimes go overboard. It is not uncommon to see PowerPoint decks with reams of pages that upper management doesn't have time to review. After the first 15-20 pages, they're saturated. Teams developing those "Tier I" KPIs with their underlying "Tier IIs" and those underlying "Tier IIIs" are not really adding value to the business. All they're doing is re-verifying what is already known. What is the true value added of all of that effort? While all of this analysis is on-going, the resource time needed to actually execute problem-solving can be starved. Simply put, Lean is flow. And that flow includes both the transformation of the products (things) as well as the transformation of the data (information). Developing and executing the Lean management of the transitional functions of a finance team are well described in Maskell's and Baggesley's book *Practical Lean Accounting: A Proven System for Measuring and Managing the Lean Enterprise.*

The financial and governance teams of an organization must ensure that the development of the supporting tactics enable the systematic reduction of their costs to optimize the performance of the business. Chad and Debra Smith's article "The Importance of Flow and why we Fail so Miserably at it" describes these opportunities well.

KPIs create visibility and accountably. Having too many of them creates waste and an inability to fully leverage one's human resources to actually execute the changes required to improve performance. Identify the key leading and lagging KPIs that are needed. Use the Hoshin Kanri to connect the KPIs from the business strategies to the tactics and stick to the vital few. Develop an introspective review of the transitional processes and make them more efficient.

Becoming Lean... Across the Enterprise

So where does one start? How does an organization leverage the functional linkage from the strategic plan to the functional teams and enable Lean to happen concurrently? There are hundreds of businesses that have been successful and there are simply too many books to cite. Each of the models out there has worked for someone. What will be the approach methodology that will work for your business?

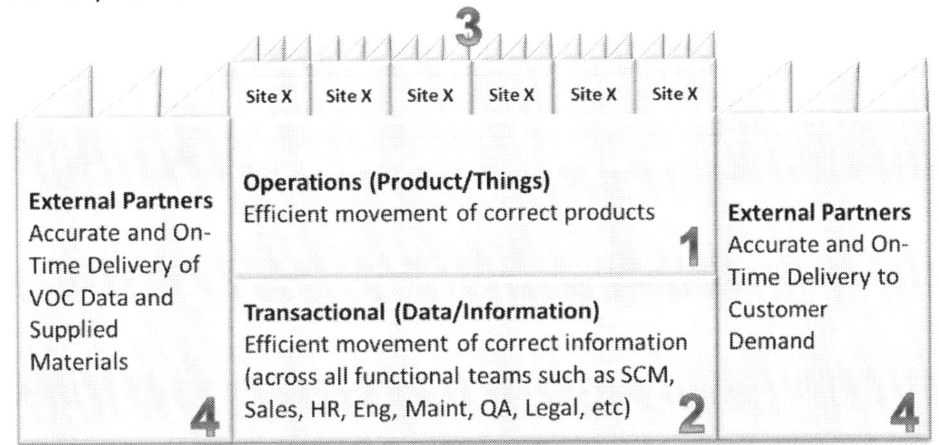

Figure 2.12

As an organization works across its Hoshin Kanri through an integrated approach with its functional teams, a phased approach shown in **Figure 2.12** to Lean generally evolves:

a. **Lean "Operations"**

Often called "Lean Manufacturing," it is the focus on the core value streams of the organization's manufacturing and/or operations. This is the improvement of the "what" that is being delivered whether it is a product or process. Lean Operations is the *effective* and *efficient* flow of this Product/Process. This is often where "**LEAN**" gets its bad wrap... that *Less **E**mployees **A**re **N**eeded*. Let's be real... **it's true**! Lean is simply about flow. If you are able to flow your products more quickly through the processes and reduce lead time, then the productivity WILL increase AND the Cost Per Unit (CPU) WILL decrease. This results in the need for LESS people to perform THOSE types of activities.

Now let's get to the meat of the matter. **Lean** is about flow (now note the *lower case letters*) AND it's also about growth. As personnel are freed up from their non-value add and/or inefficient activities, they should be re-deployable within the organization. Accountants can get whiplash with that statement, so hold tight... There *is* a business need for these personnel; AND the CPU does go down. There are generally two forms of proactive redeployment:

1. **Attrition.** Some associates will simply not be able to withstand the systemic changes that come with Lean processes and they will opt out. As this happens, personnel can fill those natural positions enabling a net reduction. Depending on operations type, our Lean professionals have seen this range from three to five percent across the business in other organizations.

2. **Growth.** With effective Lean implementation (>50% average reduction in lead time for operational processes), the CPU is commonly reduced by 20% or more. This should enable the company to be more agile and competitive with pricing and potentially secure more business without sacrificing current margin. If the market itself is stagnant, Lean systems should enable ~50-75% reduction in core transactional process, such as Lead Time for New Product Introduction, Request For Quote, etc., enabling the ability to garner new business by being able to quote first, get it to market first and provide it at a more competitive price. If the business is still struggling on those fronts, the opportunity to partner with your customers by increasing their turns and reducing inventory across the value chain via right sized inventory markets should provide a value proposition that goes beyond just the product that is being sold. In this business' results, they average a Lead Time (LT) reduction of over 50% across more than 11 manufacturing value streams as shown in **Figure 2.13**. More detail on average lead time reduction for transactional processes will be covered in Chapter 4.

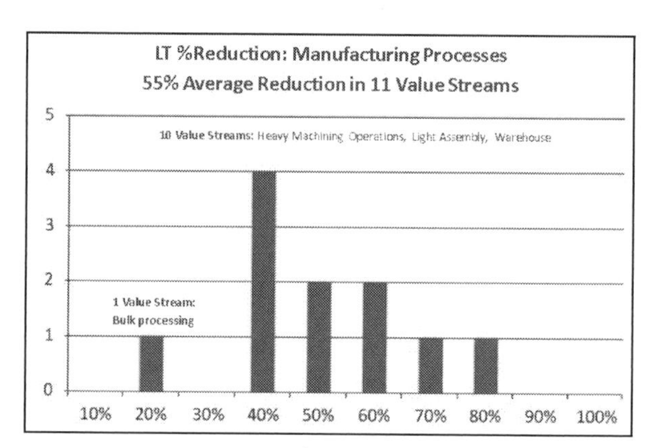

Figure 2.13

Leveraging the success of Lean can be very powerful. If, however, the organization is solely fixated on the LEAN aspect of Lean, they may likely achieve short-term gains. However, they will not likely realize long term sustainability. They also risk losing the trust of their workforce. Additionally, the business will only get so far with Lean *operations*; achieving the next levels of Lean in the business could be culturally challenging to achieve.

b. **Lean in Transactional Systems: The effective and Efficient Flow of Data/Information (Transactional Lean).**

The core value streams of the business can only be optimized so much as they are typically dependent on the accurate and timely delivery of data from outside of their span of control. Within the business, these support functions typically include engineering, supply chain, quality, environmental health and safety, sales and marketing, human resources, finance, IT, legal, etc. Typically, very few of these transactional teams provide the actual product that the customer is buying and are a necessary part of the business' support structure.

It is their responsibility to develop, manage and execute their necessary systems in such a manner that enables the most efficient (flow) and effective (quality) implementation of services. It is not uncommon to see dramatic reductions in lead time (>50-75%) as shown in **Figure 2.14** along with significant reduction in errors and rework. The E-VSMs are fully transactional and support the effective movement of product through the organization where the T-VSMs are functionally based.

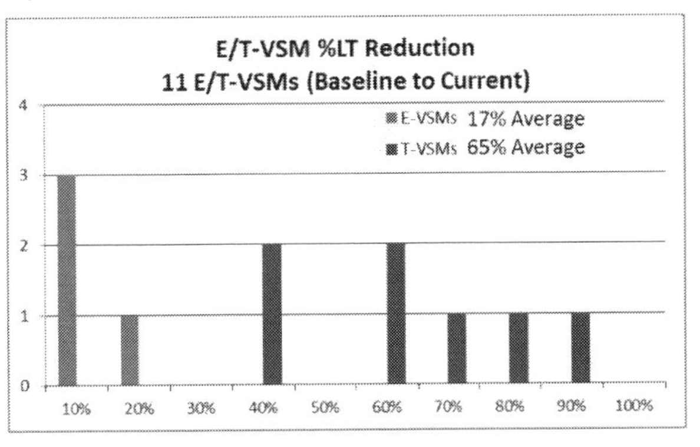

Figure 2.14

Organizations focusing on Lean "operations" often neglect taking it to this next level. A potential trip hazard is if LEAN operations are utilized. When the leadership team's language focuses on "belly buttons or toe tags," their focus is not effectively on the improvement of the business. Transactional team members are often a part of the LEAN "execution." They can translate what the impact is to them when or if the focus is shifted upward. Depending on the cultural tone set at operations, this could very well make or break the effectiveness of Lean execution across the rest of the business.

In five companies, similar rates of three to five percent of attrition were observed for these transactional roles as some personnel struggled with the systemic changes needed in a Lean environment. As a result, the same opportunities for attrition, replacement and growth fulfillment exist at this level of the business. However, there can be an additional opportunity. We've heard of some businesses using redirected personnel as ad-hoc problem solving teams until they are placed in permanent roles. This powerful "self-funding" method can bring significant savings to the bottom line while simultaneously cross-training these personnel to be effective in a variety of roles. One business used redirected personnel to replace third party contractors in various capacities. Generally, when attrition or growth openings occurred, personnel from these pools were sourced from within. Their bottom line impact well exceeded the expense of their "temporary" retention until a position was available. In one business, no new personnel were hired over a three year timeframe while the business both grew in size and market share.

From a transactional Lean perspective, many specialties are now arising as shown in the book covers in **Figure 2.15**. There is *Practical Lean Accounting* (Maskell), *Lean Sales and Marketing* (Asefeso), Lean Engineering via *Mastering Lean Product Development* (Mascitelli),

Lean Supply Chain & Logistics Management (Myerson), *Lean HR* (Lay), etc. These resources should remind serious leadership teams that Lean is not just in operations.

There are many Lean "tools" to leverage at the operations level. These often translate equally well into transactional processes. Some examples are: Value Stream Mapping, 5S Analysis, kanban, Poka Yoke, etc. One that is not as commonly translated, but is equally powerful, is the ability to apply Standard Work (StdW) analysis. It's easy to grab an Industrial Engineer (IE) with a stop watch to determine the best way for an operator to load piece X into machine Y and then do steps A-Z. Imagine translating this type of breakdown to material

Figure 2.15

handlers, maintenance personnel, financial personnel managing credit/debit memos, supplier quality engineers, software engineers, laboratory technicians, etc. Where appropriate, conducting StdW analysis on indirect labor positions can be very powerful. Personnel tied to operations add value to the business and can be related to the variable fluctuation of sales. Those in transactional positions need to optimize their processes as much as feasible to maximize margin potential. The basic tenets of Lean apply across the business.

We now need to ask ourselves a critical "chicken or egg" question: ***Should an organization start Lean via the traditional Lean <u>operations</u> route, or should they start at the <u>transactional</u> levels?*** Is it possible to successfully start with the transactionals first? We've not yet heard of an organization that has started their Lean journey from this perspective, and we'd be interested in hearing from anyone who has. Imagine the change in perspective of how Lean could be deployed if it is first started "upstairs." How fast would the rest of the business follow through with execution when the transactional process flows are already effective and efficient. Or, would it never get off the ground as the office personnel try to figure out how it applies to them? The "floor" can only be as optimized as the transactional processes will enable them to be. While it is easy to get Lean value stream numbers, such as lead time, productivity and CPU when dealing with pieces and parts, it would take a very selfless and servant-based organization to look at itself in the mirror and say, "Let's ***really*** start at the top."

In the demonstrated transactional StdW sample, ~50 professionals were evaluated for how they spent their time. **Figure 2.16** shows the time distribution of these professionals. About 10% of it was purely due to rework and waiting time for late and/or incorrect information. Another ~12 percent was spent in meetings that were not deemed as value added or productive. Imagine our reaction if our direct labor teams spent a full 10 percent of their time on rework and waiting... and another 12 percent sitting in meetings not being effective? We'd have a full court press on resolving the situation. If the transactional team focuses initially only on that portion of their system wastes, then the organization has an opportunity to re-deploy 10% of ***its*** staff to other parts of the business.

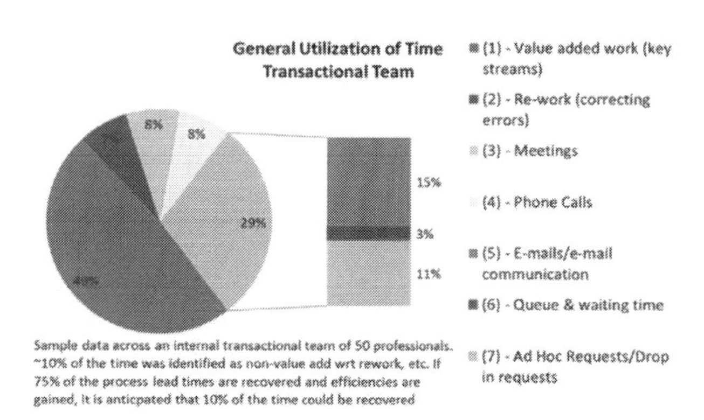

Figure 2.16

c. **Align Key Value Streams Across the Business Enterprise: Both Product & Transactional**

At this phase of the Lean journey, it is anticipated that the operational pieces are well in hand. We consider this to be an "inside out" approach. As the core value streams are optimized in flow and quality, the external processes can then be optimized to support them more effectively (E-VSMs are described in more detail in Chapter 6). For larger, multi-site organizations that are vertically integrated, shipping Inter-Company (I/C) is a common occurrence. Managing those system orders efficiently (flow) and effectively (quality) will prescribe the overall limitations of the business. By focusing on the product and process transactions across the organization, the business will take its next great leap in reducing its LT. Again, the adage "time is money" will become even more transparent as the egregious wastes in the system are exposed. Previously in **Figure 2.13**, an average of 17 percent additional LT was removed at this secondary level, subsequently reducing working capital.

Jonathan Chong's article "Enterprise-Wide Value Stream Mapping: Create a Vision of Your Company That Really Puts Your Customers First" describes how a business can look at itself from that 50,000 foot level. He identifies five main components in an Enterprise VSM:

1. Transactional: The **marketing and sales value stream.**
2. Transactional: A **product development value stream**. As an example, the automotive industry has a very well defined model issued by the AIAG. It is called APQP. Product planning is commonly executed across a "staged gate approach." APQP identifies five stages (Other product development systems identify six, seven or more gates). Each of

these gates should be considered when an organization is conducting its value stream analysis of their product development system. Some key gate considerations (per AIAG) are:

i. Concept/Initiation approval of the new product or process. Typically based on Voice of the Customer and/or marketing analysis.
ii. Program approval. Feasibility evaluation where the organization determines the viability of providing the product or process in a manner which meets strategic objectives.
iii. Prototype development. Creating and testing the new product and/or process. This includes both the physical creation of the product/process but also the manufacturing or system development across which the product/process will be performed.
iv. Product/Process launch. The issuance of the new product or process. The business' change process should also be included. This is typically an engineering transactional function.

3. Operations: An **operations value stream.**
4. Transactional: The **IT infrastructure** showing data flow.
5. Transactional: The **supply chain (and logistics) value stream**. This can include several perspectives:
   i. The planning management of receiving goods including the development of pull signals via kanban markets.
   ii. The planning management of Outside Service Providers (OSPs) who add value to the product outside of the "four walls" of the business.
   iii. The planning and coordination of shipping and logistics of the finished goods and services.

Figure 2.17 shows a high-level concept sketch of transactional processes, operations and supply chain. A formal VSM would have much more information. The intent here is to show the complexities across multiple sites from an I/C perspective. After the "within the four-walls" VSMs were conducted, Accuride evolved into the E-VSM within the "four-walls of the business," and millions more in working capital were removed within less than half a year. **Figure 2.18** shows the impact on the inventory.

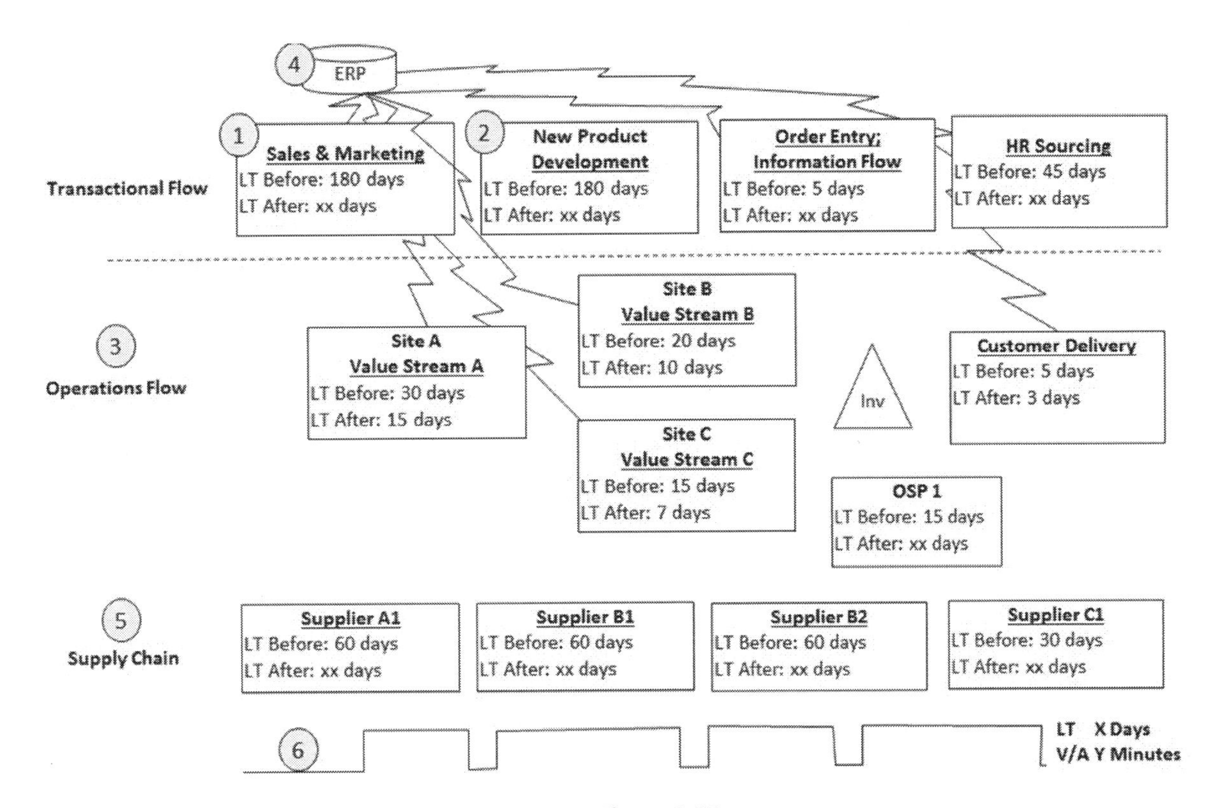

Figure 2.17

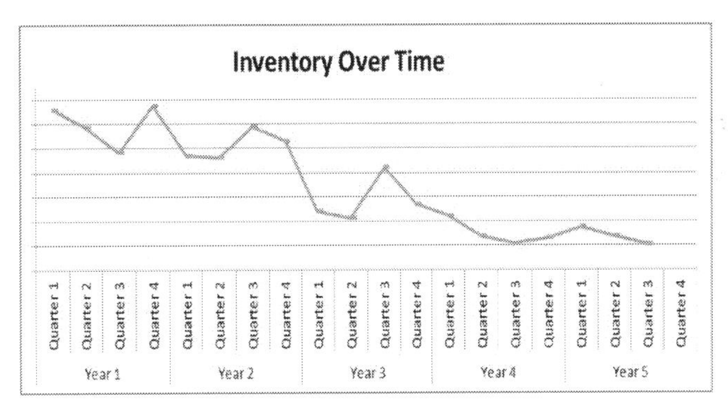

Figure 2.18

Accuride further adapted this model into a "3D" version. The E-VSMs were evaluated at Runner, Repeater and Stranger levels giving even more opportunistic insight into inventory management. These are also called "ABC" items.

Today, organizations seem to focus on
their operational value streams first. By doing this, Accuride achieved a ~50 percent average reduction in LT from within the "four walls" of each of its various sites. This was also conducted simultaneously as the local personnel were effectively cross-trained in the methodologies and tied to performance reviews by a Steering Committee (Accuride calls theirs a QLMS Council) and quarterly Lean status reviews. With this parallel site approach, the velocity of execution within a business unit with six to ten sites can accelerate rapidly. We have found the utilization of the X-Matrix to be a key-enabling component.

There are often rapid and sizeable returns within the "four walls" of the various sites. As the sites continue on this journey, the teams often recognize the need for the transactionals to catch up, traditionally in sections two, three, four and five as shown in **Figure 2.17**. As the company expands its focus, section one gets incorporated into the fold along with the I/C product and process logistics. It is critical to note that the quality management system needs to be in good order for this to occur or the flow cannot occur without frequent interruptions.

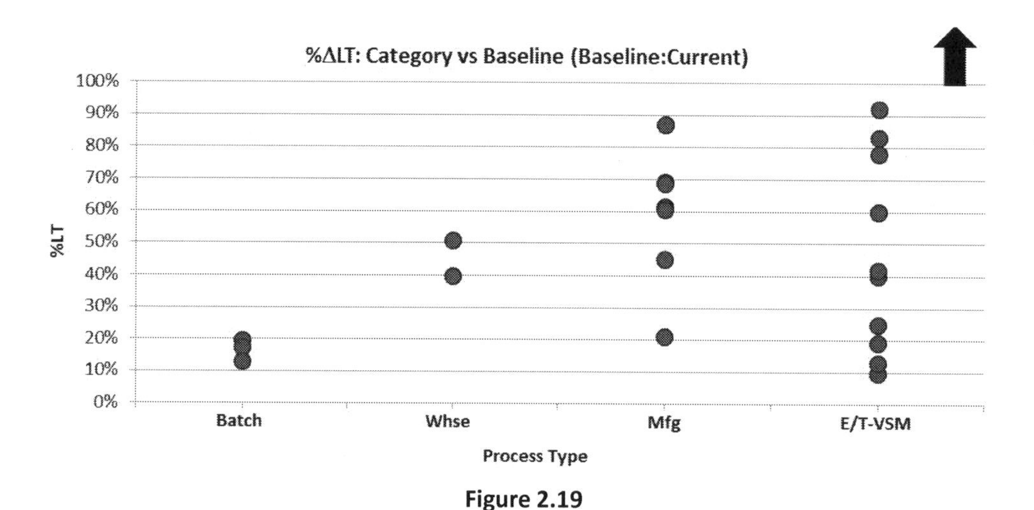

Figure 2.19

Figure 2.19 shows the impact on the LT across various process types. Accuride achieved a 55 percent average reduction in LT across the sites in all of these processes. *When the E-VSMs were executed, an additional 3.9 percent of COGS was removed from across the entire value stream.* The more complex the process, the more LT seems to be able to be impacted. COGS is a critical metric largely used to describe a product's value. It has three basic components: Materia

produce the product p

Direct Labor

Overhead Costs

Raw Materials & Components

Margin

Figure 2.20

50-50-20

costs and Taxes. **Figure 2.20** shows the COGS details with Direct Labor (3-5 percent of COGS), Overhead (32-25 percent COGS) and Material (65-70 percent of COGS). In North America, and depending on product/process type, it is not uncommon for materials to be the largest component of a product's COGS. Oftentimes, it can be as much as 60-70 percent. With materials being such a high part of COGS, it seems that Lean execution teams can be misguided when placing such an emphasis on the direct labor component first.

Typically, the next highest contributor is overhead. Upper management's driven focus on LEAN can be very misplaced... most of the money is tied up in the raw materials, work in process and finished goods inventory while moving it around from place to place within the four-walls of a site. This is not talking about Low Cost Country (LCC) sourcing or beating suppliers into cost submission; it is purely about reducing the amount of what is on-hand and only buying what is truly needed.

d. **Incorporate Partners: Customers, Supply Chain, Contract Services, etc. (Largely Transactional)**

Continuing on that path of "inside out," the fourth phase of focus includes key partners external to the organization. Whether or not the organization begins with operations or transactionals, working with the external entities during the last phase makes the most sense. The proverbial house should be in order first before going outside. There are a few large entities to consider in the scope of external partners. These would typically include:

1. **Customers:** Front-end transactional processes, such as sales and forecast planning, order entry acknowledgements, order entry processing, etc. Back-end transactional processes, such as logistics delivery, sequencing, billing and the inevitable debit memo processing.
2. **OSPs:** Supporting their front-end transactional processes, such as sales and forecast planning, order entry acknowledgements, order entry processing, etc. Understanding their back-end process needs, such as logistics delivery, sequencing and billing.
3. **Suppliers:** Supporting their front-end transactional processes, such as sales and forecast planning, order entry acknowledgements, order entry processing, etc. Understanding their back-end process needs, such as logistics delivery, sequencing and billing.
4. **Compliance Organizations:** Financial, Quality, Supply Chain and EHS systems typically have some form of compliance criteria in which internal and/or third party support is needed. Examples would be Sarbannes Oxley (SOX), General Agreement on Tariffs and Trade (GATT), Committee Of Sponsoring Organizations of the Treadway Commission (COSO), third Party Quality systems, such as ISO 9001, third Party EHS systems, such as ISO 14001, Local state environmental and regulatory requirements, FDA Regulations, etc.

Businesses can proactively partner with these organizations to both optimize these processes and reduce the amount of physical third party support that is actually provided. There can be fiscally impactful results in both effectiveness (quality) and efficiency (flow).

It can sometimes be very challenging to get customer partners to the table. Some customers simply believe that a warehouse of goods is the best way to do business. Helping them understand the value proposition of Lean can take time and persistence.

Most businesses start with their suppliers and OSPs. From there they grow to incorporate the compliance organizations. They share results to demonstrate their newfound agility. As Lean becomes more and more recognized for its benefits, some customers create a pull with their suppliers and request partnered Lean system approaches. Leveraging these opportunities can be beneficial to both teams.

Results Across the Value Streams

As organizations decide how to develop their Lean systems, they need to evaluate whether or not they will leverage third party support or do it themselves. There are pros and cons to both approaches. Regardless, the Lean journey is hopefully kicked off.

To manage the business' overall strategy, the development of a Hoshin Kanri that links the business' key strategies to that of the organization's functions can be critical in enabling their success. The Hoshin Kanri process methodically destroys silos and increases employee engagement. By creating these linkages, the overall business experiences fewer initiatives and develops surgically precise tactics that are optimized to advance the company's strategies.

As the strategies and tactics are developed, the KPIs are then determined to monitor progress. Developing leading and lagging indicators, usually in a 2-to-1 ratio of lead to lag, can enable an organization to be more "pre-emptive" rather than reactive to end results. The Hoshin Kanri methodology also helps to strip out redundant or correlating metrics, thereby enabling a very streamlined metrics system. Utilize the basic tenets of process control to enable focus on variation reduction and basic process improvement.

With the Hoshin Kanri and KPIs in place, execute the tactics. Manage the business enterprise through the four segments of Lean systems in a typical enterprise (See **Figure 2.11**):

1. **Operations:** The efficient movement of the correct things.
2. **Transactional:** The efficient movement of the correct data or information.
3. **Enterprise:** The efficient movement of the correct things and information across the business entity.

4. **External Partners:** The transactional effectiveness with customers, suppliers, outside providers and compliance organizations.

By managing these systems, a business professional can more objectively develop their own process to better enable their organization's success. Depending on the type of key value stream(s) the organization has, and after working across dozens of value streams, it is not uncommon to see the following types of results:

- >60 percent average reduction in LT for operations based processes. This can vary depending upon the process type: Heavy assembly/component throughput, large scale machining, capital equipment building, etc. The more complex and human controlled the process, the more the LT can generally be reduced.
- >37 percent average reduction in LT for batch-based systems, such as painting, heat-treating, bulk recipe batches, etc., where batches of products are produced simultaneously.
- >47 percent average reduction in LT for transactional and enterprise-based flow based processes. These had the most amount of variation depending upon the type of process complexity.

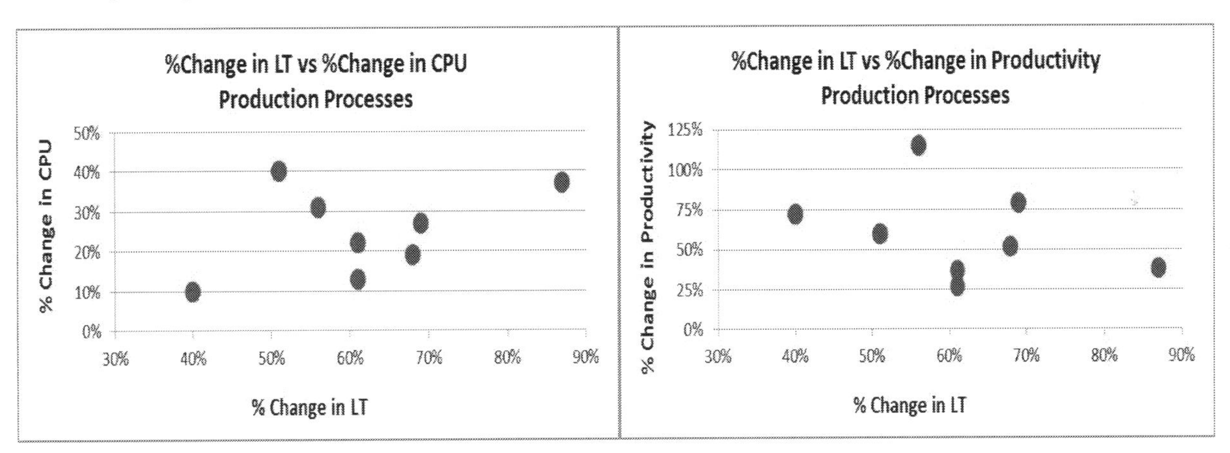

Figure 2.21 **Figure 2.22**

Again and again, we recognize that time is money. **Figure 2.21** and **Figure 2.22** summarize the reduction in lead time and its impact on both CPU and productivity. While the changes in both productivity and CPU are significant, *there does not seem to be an overall correlation where "more lead time out" gets to an ever higher and higher CPU or productivity performance level*. Based on these diverse processes of batch, warehouse and varied types of production, there does seem to be a tipping point where **at the 40-50 percent of LT reduction (from baseline), the productivity and CPU has become somewhat optimized**. At this point, it would be time for process redesign.

These types of process results create an agile and high-growth potential atmosphere. The business can achieve several competitive advantages from time to market, productivity, higher margins and market share. As these teams continue on these journeys, there is a possible “secret sauce” formula related to lead time that a business might consider:

50%-50%-20%©™

Where an average 50% reduction in process lead time can increase the productivity by more than 50% and reduce the CPU by an average of 20%.

As more processes are incorporated and evaluated for their performance impact, the observed relationships may change. Even so, these guidelines may provide a starting point for an operations based organization.

As our sites hit these milestones, they are now positioning themselves to do it again.

Will Rogers probably said it best. “Even if you are on the right track, you will still get run over if you just sit there.”

Key Takeaways

1. Personnel with clear goals that connect their actions to that of the overall business generally have higher levels of engagement.
2. Benchmarking is a process that forces us to look at our organization through that different lens.
3. Hoshin Kanri is a step-by-step planning process that uses an iterative matrix model to connect organizational strategies to the tasks and measurements.
4. KPIs are a blend of leading and lagging indicators that help to control the front-end so that the back-end follows.
5. To get the full value of Lean, there is a need to expand the scope to look at enterprise wide processes.
6. Successful Lean implementations reduce lead times to a varying degree depending upon the process, which results in increased productivity and reduced cost per unit.

Chapter Three: Tracking Progress

How do we know how we are doing? Most of us would like to look at the financial statements, such as operating income, EBITDA, return on equity, free cash flow, etc. It is true that financial statements reflect how well we *did*. However, they provide the net financial result of **all** the actions that drive these results.

Is this the reason why we establish our businesses? If we look at Accuride's vision, mission and values (a result of strategic planning), it becomes clear that financial results are not the only reason for being in business. Accuride's vision, mission and values are:

At Accuride Corporation, what we believe and what we value drives what we do and how we do it at every business unit, every facility and every workstation. We're a business, yes, but ***we don't support a "profits at all costs" mentality****, and we don't believe that "good enough" is ever good enough when it comes to the quality of our products, the value of our services or the responsiveness of our associates.*

Our Vision: *Accuride will be the premier supplier of wheel-end system solutions to the global commercial vehicle industry.*

Our Mission: *To create and support a culture and team that provides:*

- *High value products and services to our customers.*
- *A great place to work and grow for all of our associates.*
- *Superior financial returns for our shareholders.*

Our Values: *At Accuride, we are committed to being:*

- *Ethical and respectful*
- *Customer-centric*
- *Technology and quality leaders*
- *Cost competitive*
- *Safe, progressive and inclusive*
- *Environmentally conscious*
- *Fiscally conservative*

It is clear that while profit is critical, it should not be the only driver. It is important to realize the difference between outcomes and outputs (**Figure 3.1**). Accomplishing vision and mission are the outcomes, and these are the results of certain outputs. This means there is no outcome without output. However, just because there is an output, it does not mean we are getting or will get a desired outcome.

If we want to see a significant impact on an output, first we need to understand the key inputs and then start monitoring to ensure we consistently get the desired output. Each measure must be precisely defined to monitor what is its intended outcome; or the effort is wasted. This process helps to establish a cause and effect relationship with leading and lagging indicators.

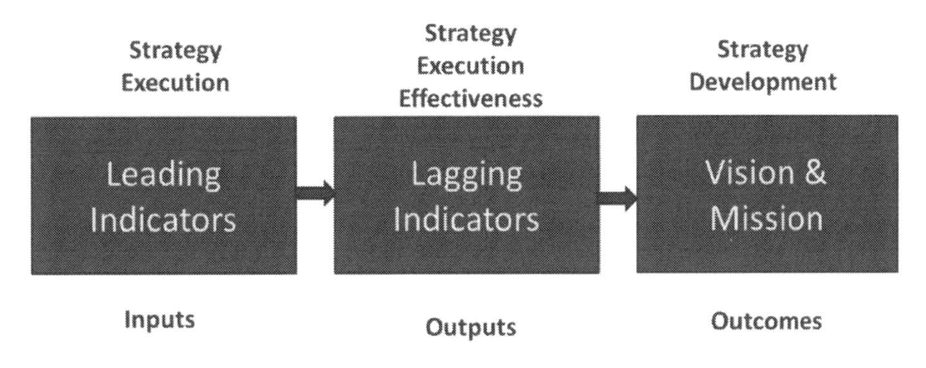

Figure 3.1

In operations, we can track many leading indicators (inputs) that drive the lagging indicators (outputs), which lead us to achieve our vision and mission (outcome). **Figure 3.2** is a sample operational lead/lag correlation table. The metrics on the left are leading indicators to the lagging indicators commonly used by external customers to measure the output performance of a QLMS. Referring to the previous Ben Franklin quote, the leading and lagging indicators should correlate.

There are other types of metric correlation tables for the other functions, such as HR, IT, Finance, Supply Chain, Sales, etc.

Sample Leading Indicator Metrics	Lagging Metrics			
	OTD	PPM	COPE	$Conversion
OEE: Operational Equipment Effectiveness	+	+	+	+
RTY: Rolled Throughput Yield	+	+	+	+
%Cpk: Processes Capable of Meeting >=1.33	+	+	+	+
MTBF: Mean Time Between Failure	+		+	+
C/A %Repeat: Repeating Complaints		+	+	
Training Depth: Personnel Skill Levels		+	+	
Linearity: Production Schedule Attainment	+		+	+
C/A TTC: Time to Close Corrective Actions		+	+	+

Figure 3.2

The X-Matrix (Kanri) process is very effective at cross-correlating which KPIs (both leading and lagging) should be monitored within the business. They will tie to both the strategies and the tactics. Here are some of the key considerations when developing KPIs:

1. **Don't Over-Measure. Simplify Inputs.** Capture key inputs that focus on the most important causes.
2. **Eliminate Correlations.** Re-validating the same information reduces the amount of resources available to focus on and fix the issues at hand.
3. **Trend Review.** Review the results for performance graphically to identify trends. Trend review minimizes over and under-reaction to single data points. You can see a sense of movement.
4. **Keep Your Metrics to a Vital Few.** Six to eight lagging indicators is the norm for a mature Lean business with two to three leading indicators per lag. In all, 20-30 KPIs should cover the business for all interested parties (Stakeholders, Employees, Customers, Suppliers and Community).

Again, the Hoshin Kanri should determine the KPIs. When those are identified, targets should be established to meet the business needs. Accuride manages ~21 QLMS based KPIs at different frequencies as shown in **Figure 3.3**.

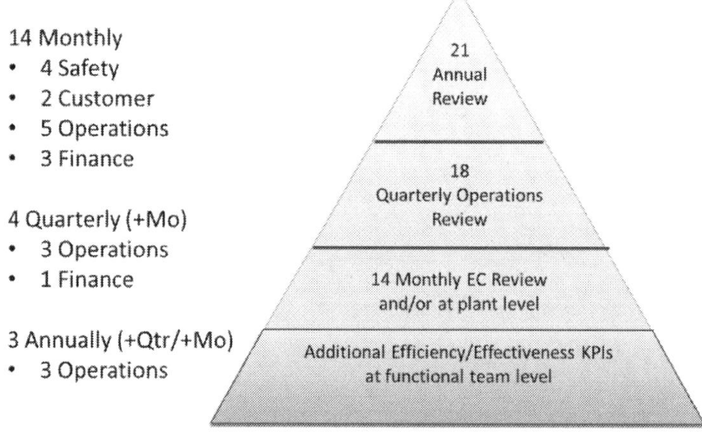

Figure 3.3

There are many "cross-over" KPIs in here for Supply Chain, Engineering and other functions. The Finance team manages a sub-set of leading and lagging indicators as well.

Here is the brief description of some more relevant operations-based KPIs tracked by the QLMS teams at Accuride with selected charts for historical trends from the original baselines.

Safety KPIs (Leading)

Near Miss. Near miss is tracked monthly and is a rather broad term, which incorporate multiple levels up until an injury, or recordable, is incurred. These are reported by any Accuride employee across the corporation at any location. This is a leading indicator, which if controlled, will help to reduce the Total Recordable Incident Rate (TRIR) and Lost Work Case Incident Rate (LWCIR).

Figure 3.4 shows Heinrich's Pyramid of incidents. Sites are encouraged to report any and all potentials. A corporate-wide EHS council meets monthly to review both Lessons Learned and Learning To See opportunities.

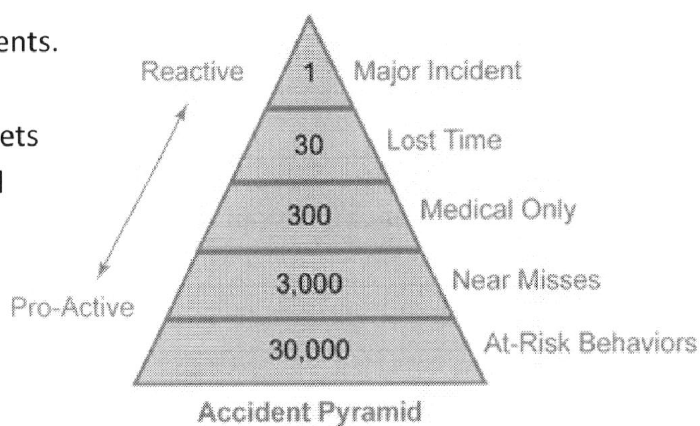

Figure 3.4

Safety KPIs (Lagging)

1. **TRIR:** TRIR is tracked monthly and is used to estimate the number of total recordable incidents that would be expected to occur among a group of 100 employees. It is a lagging indicator because it is after the fact. It is a measure of the effectiveness of the safety program (It could be a leading indicator for the KPI below). The TRIR calculation is:

   TRIR = (Number of incidents experienced by employees in the current month) x 200,000
   (Non-OT hours worked in current month + OT hours worked in current month)

   A simple way to remember the hours worked is to account for all "belly-buttons." Anyone and everyone who was there is to be accounted for. **Figure 3.5** shows the TRIR trend at Accuride from its baseline.

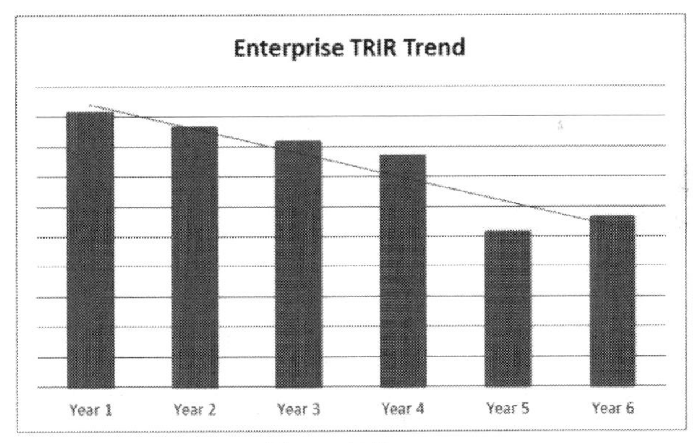

Figure 3.5

2. **LWCIR:** LWCIR, tracked monthly, is intended to estimate the number of total lost time incidents that would be expected to occur amongst a group of 100 employees in one year. It is a lagging indicator because it is tracked after the fact. It measures the safety program effectiveness and is calculated as:

   LWCIR = (Number of Lost Time Incidents in the current month) x 200,000
   (Non-OT hours worked in current month + OT hours worked in current month)

   Lost time incidents include results from all people at all Accuride facilities. Accuride's LWCIR is below 1.00 worldwide.

Customer KPIs (Lagging)

1. **External PPM:** Often called "0 Miles," this monthly KPI tracks the frequency of non-conforming goods reaching the customer at their build locations prior to the unit being assembled. It is measured in PPM. This is an after the fact, lagging indicator.

   External PPM = (Number of non-conforming parts received in the current month)
   (Number of parts shipped in current month)

   Non-conforming parts received by the customer include pieces returned, parts scrapped at the customer site, parts repaired at the customer site, etc. and designated as non-conforming by the customer. Non-conforming parts should be reported by producing location and by field engineering. Parts shipped are the number of production parts shipped to the customer.

2. **On Time Delivery (OTD):** OTD is tracked monthly at two levels; what the sites mutually committed to (Mutually Agreed Delivery Date (MADD)) and to the dates that the customers requested (CRD). This helps us to debug internal and external delivery issues when customer requested carriers do not show or when orders are received inside the lead time and we need to adjust market plans. These are both lagging indicators, but the CRD cannot be successful if the MADD is not at 100 percent. It is defined as:

   OTD = (Number of pieces delivered on time in the current month) x 100
   (Number of pieces due in the current month)

Figure 3.6 shows Accuride's enterprise-wide OTD trend.

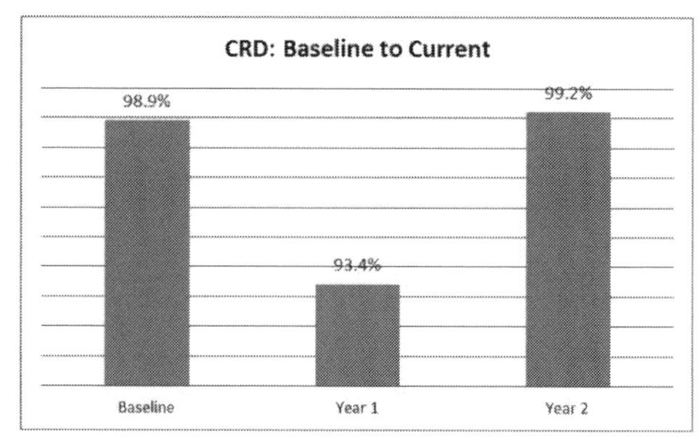

Figure 3.6

Operations KPIs (Leading)

1. **OEE:** OEE is tracked monthly. It provides an estimate of process throughput taking process capability, system effectiveness and process efficiency into consideration. It is the percent of time that the equipment is operating at its rated capacity producing quality results (See Chapter 1).

   OEE = % Planned Uptime x % Yield x % Efficiency

   % Planned Uptime = (Time Actually Spent Producing) / (Time Scheduled).
   Note: Not time availability.
   % Yield = (Quantity Good)/(Quality Produced).
   Note: Product that is reworked or salvaged is not considered "good" at its first pass and should be removed. %Yield cannot exceed 100 percent.
   % Efficiency = (Quantity of Product Produced)/(Ideal Quantity Expected). If the production rate exceeds standard, then only 100 percent should be used and the standard adjusted to reflect the improved process. % Efficiency cannot exceed 100 percent.

   Figure 3.7 shows the OEE trend at Accuride from baseline.

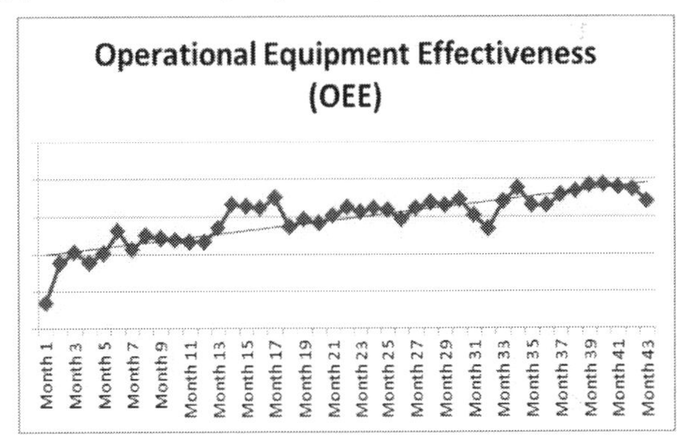

Figure 3.7

2. **Rolled Throughput Yield (RTY):** Tracked monthly, RTY measures the capability of the overall process. It iteratively multiplies the yield (right first time) from each step of the process. It is one of the largest and most powerful indicators of margin erosion.

   RTY = %Yield @ Step 1 x %Yield @ Step 2 x %Yield @ Step 3 x %Yield @ Step X...

   Process steps are determined independently at each location. It is intended that about 80 percent of the revenue stream be captured through this metric. **Figure 3.8** shows the RTY trend at Accuride.

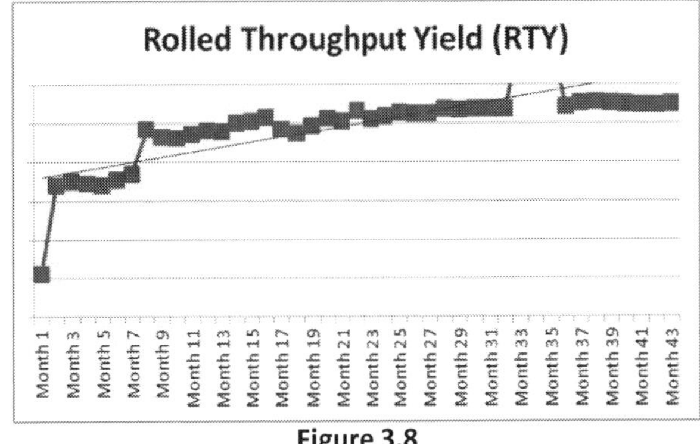

Figure 3.8

3. **Percent Pull to Plan:** Tracked monthly, %Pull measures the ability of the facility to meet the kanban markets. It is a measure of Heijunka and level loading. %Pull is a leading indicator of production efficiency.

   %Pull to Plan = (Quantity of units planned to have been pulled in the current month) / (Total Units actually pulled that month)

   %Pull to Plan is applied to at least 75-80 percent of the site's production volume.

4. **%Key Process Indicator Variables (KPIVs) with Cpk>=1.33:** This Leading Indicator is tracked quarterly. The %KPIV monitors the health of a site with respect to its ability to meet all critical features effectively. A comparator to a Cpk of 1.67 is used for safety-based features.

   %KPIV = (Number of KPIVs with a Cpk >= 1.33)/(Total number of KPIVs managed at the site)

   KPIV lists are site dependent, but the performance criteria are not. The intent is to cover KPIVs for at least 80 percent of the revenue stream. Accuride is considered Benchmark in its KPIV performance levels (**Figure 3.9**).

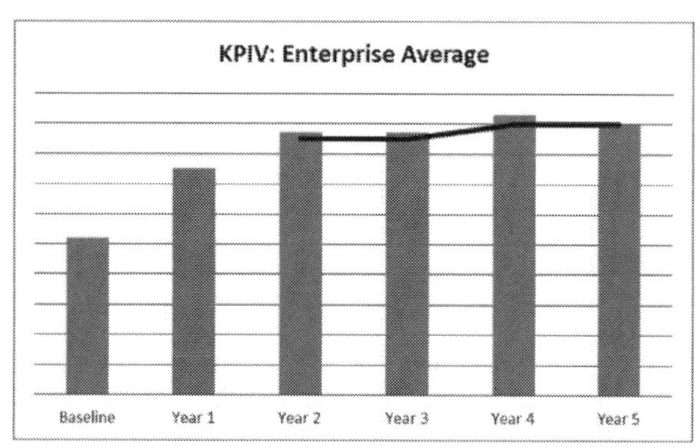

Figure 3.9

5. **Mean Time Between Failures (MTBF):** Tracked quarterly, MTBF is a measure of the stability and effectiveness of key process equipment. It is a leading indicator of effectiveness; MTBF is the average of the elapsed time between the failures of a system during planned operation.

   MTBF = (Total Run Time Hours)/(Unplanned Downtime Occurrences)

   Key assets are identified and will vary from site to site. In addition to managing equipment, which runs 80 percent of facility revenue, additional equipment is selected based upon capacity and redundancy or risk to business if the unit becomes unavailable.

   Accuride's MTBF trend is shown in **Figure 3.10**.

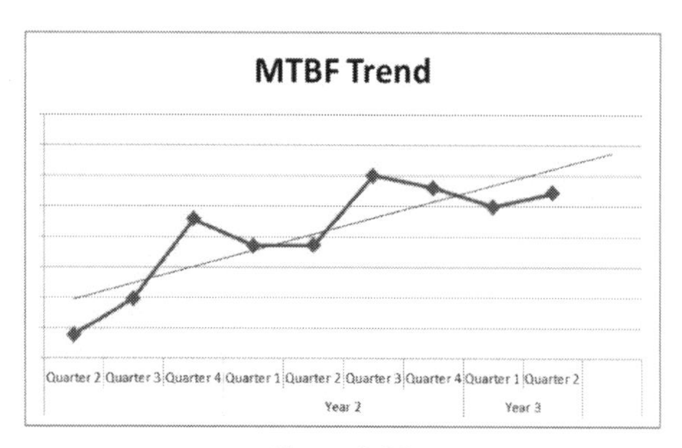

Figure 3.10

6. **Bench Strength (Bloom's Taxonomy):** Tracked annually, bench strength measures the experience and effectiveness of the skillset of QLMS personnel on Accuride's Council.

This is a leading indicator of efficiency. Blooms uses a one to six scoring across the QLMS skillsets required. A score of one to two is at a beginner level, a three to four is "training wheels on" and five to six is "training wheels off." An average of the scores across the individual skillsets is taken and the overall corporate average is tracked.

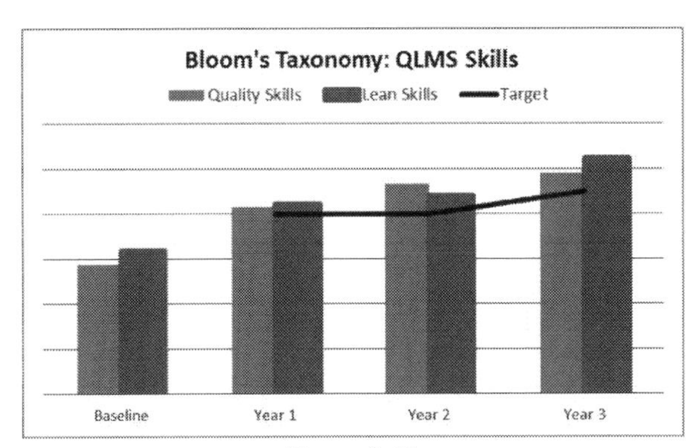

Figure 3.11

QLMS Skillsets in the Lean and quality sciences include an alphabet soup of key tools such as: Cp/Cpk, MSA/AAA, SPC, DOE, FMEA/CP/DCP, StdW, 8D/CAPA, ISO/TS/IATF, LPA Auditing, VSM, PFEP/TTR, TPM, CFM/Pull, etc. (If interested, Accuride would be happy to provide additional details) (**Figure 3.11**).

Operations KPIs (Lagging)

1. **Units per Man Hour Worked (U/MHW):** Tracked monthly, U/MHW is a productivity measure. It is a lagging indicator of efficiency.

   U/MHW = (Equivalent Units in the timeframe)/(Non-OT + OT Hours worked in the timeframe)

   This standard metric is a key form of evaluation at Accuride shown in **Figure 3.12**.

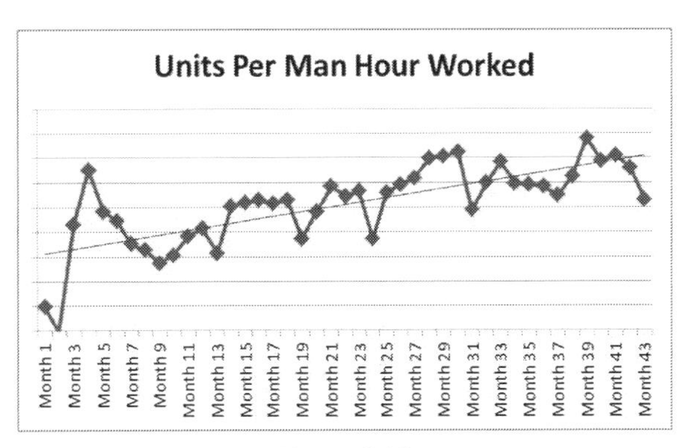

Figure 3.12

2. **QLMS Score:** There are many ways to internally assess the health of one's QLMS. Accuride uses Bill Waddell's nifty self-assessment tool with 100 questions. This handles the full gamut of QLMS and the other functional teams such as Sales, HR, EHS, Finance, Production Control and more.

Finance KPIs (Lagging)

1. **Days Inventory On Hand (DIOH):** This is one of the most critical Lean KPIs that shows the effect of the processes in play. It is tracked monthly and is a lagging indicator.

   $$\textbf{DIOH} = \frac{\text{(Prior Month Ending Inventory + Current Month Ending Inventory) / 2}}{\text{(Monthly Cost of Goods Sold/30)}}$$

   Accuride's DIOH trend is shown in the **Figure 3.13** and is again reviewed in Chapter 9.

2. **COPE:** COPE is heavily covered in Chapter 1 because a business cannot become Lean if it is incurring extensive quality losses. The COPE graph shown in **Figure 1.10** has experienced double-digit %YOY reductions for four consecutive years. There is still room for improvement.

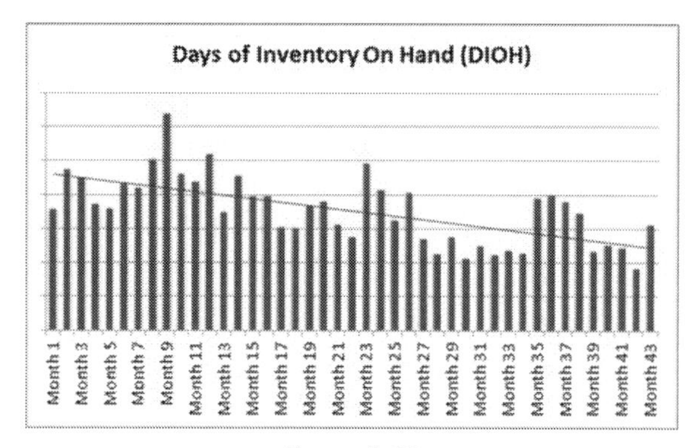

Figure 3.13

   COPE tracks each site's top four to six unplanned losses. The intent is to understand the top 80 percent of the unplanned losses. These may include: Scrap, Rework/Repair, Inspection, Excess & Obsolescence, Unplanned Premium Freight, Unplanned OT, Unplanned Loss of Capacity (empty molds), etc. When compared to COGS, COPE acts as the potential ***entitlement*** of margin recovery.

KPIs are routinely tracked. Accuride regularly reviews and adjusts them as the business matures. Some KPIs have been retired and others added. The mix has remained in the 20 range.

Key Takeaways

1. It is important for an organization to know how well it is performing. It is even more important to ensure that we are tracking appropriate KPIs to achieve the intended business outcomes.
2. To measure effectively, we need to simplify inputs, eliminate correlations, review and act on the trends of a manageable number of KPIs.
3. At Accuride, we track 21 QLMS KPIs at differing frequencies – 14 monthly, 4 quarterly and 3 annually.
4. The KPIs are classified into four categories – Safety, Customer, Operations and Finance.
5. Operations category, which is where all the action happens, generally has two leading indicators for each of the lagging indicators.
6. New KPIs may be added and the old KPIs may be retired as the business needs change.

Chapter Four: Mapping Flows

The backbone for getting effective flow is to understand every step in the process. There are generally two process types: Transactional and Operational. Both need to be performed effectively (right the first time) and efficiently (with minimal waste). This section will <u>not</u> go into how to construct the various types of process maps; there are simply too many excellent references that already do that. One of the best basic references is *Learning to See* by Shook and Rother. Other strong process mapping references include *Value Stream Mapping* by Martin and Osterling; *Value Stream Management* by Tapping, Luyster and Shuker; *Continuous Improvement Tools: Value Stream Mapping* by Kirov; and *Value Stream Mapping for Lean Development* by Locher (**Figure 4.1**). This section, though, will review the various types of mapping processes that Accuride has used and continues to use on its Lean journey as well as the context within which they are used.

Figure 4.1

Production Maps: Value Added Processes

Production processes are used to deliver products and/or services. These are generally the value added component of the business, which generates the revenue for the organization. As discussed previously, **Figure 4.2** shows Lean implementation across sites, processes and partners.

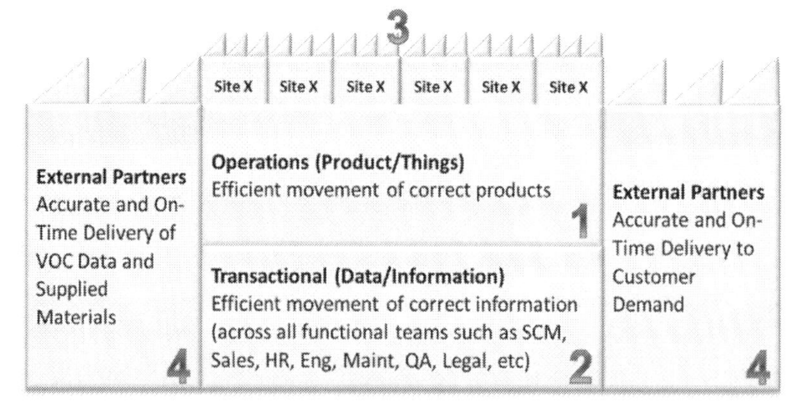

Figure 4.2

Many organizations often begin with "Lean Manufacturing" within the "four walls of a site." Constructing an effective process map can be challenging if the organization is not open to hearing the truth about the losses and illogical findings that will become readily apparent. If the leadership team is defensive, punitive and not open to objectively viewing the inherent problems that exist, then the organization is simply not ready for this basic tool. The personnel conducting the review will not show how "ugly the baby really is."

However, if the leadership can really handle the feedback, the basic "Value Stream Map" for production process flow can literally recover millions in working capital in a very short timeframe. There are several immediate benefits of mapping out a production process:

- **Quality Losses are Immediately Highlighted:** Yield/Scrap, Rolled Throughput Yield (iterative losses across the flow), Process Capability, etc. are brought to the forefront. These losses must be effectively managed in order for rapid flow to occur. If speed (flow) is attempted without rectifying these losses, then margin erosion will continue via sideline rework, scrap recovery, overtime, premium freight, material shrink, E&O and other losses. The process simply cannot go fast if things are falling by the wayside.

  Finance teams often believe that there is a value to scrap, especially if the scrap can be sold for some amount (Common in the metals and food industries). This can cause teams to be unaware of the total losses that are being incurred. If the scrap products are afforded a book value, then other losses, to be accurate, should also be included. These could include double/triple handling, human and equipment resources to manage the waste products and processes, etc. There is the loss of capacity, which is typically counted once, but what about when the replacement unit needs to be manufactured... it is consuming capacity for making another item. In addition, there is the time in counting, calculating and reviewing the scrap volume itself along with the maintenance of "revenue." Take note; if the scrap is dependent on market pricing and the value goes up, then the perceived scrap in the site goes down if compared to COGS. The opposite is also true. Therefore, if teams only track "quality" based on scrap, they may react to trends that do not actually depict the process and fail to fully understand the total business losses that are being incurred (There is more detail on this in Chapter 9). When evaluating process yield, Accuride's does not adjust for scrap value.

- **Inventory Amounts in Raw Materials, Work In Process (WIP) and Finished Goods (FG) are Highlighted:** Inventory is a counter-intuitive loss. The more there is, the slower the lead time and the greater the DIOH. The longer a changeover takes, the more the site is ingrained into thinking that they need higher levels of inventory. Key management tools to make materials flow are the leveraging of the PFEP and TTR (*Making Materials Flow* by

Harris, Harris & Wilson, **Figure 4.3**) to minimize the amount of materials needed throughout the process flow while simultaneously increasing speed and reducing DIOH. They hold a powerful in-situ three-day session for teams serious about impacting their working capital.

Figure 4.3

- **Labor Inefficiencies are Outlined:** Via the evaluation of who is doing what, how often and when, hundreds of hours of opportunity can be recovered. This balance evaluation is conducted in parallel with a VSM and is covered in Chapter 5. Whenever possible, excess labor that is identified should be redeployed by backfilling natural attrition and/or supporting business growth. OC Tanner described this as absolute criteria for their Lean journey.

As described in Chapter 2, if the version of Lean is LEAN, personnel support will be very difficult to attain. The intent of Lean is to increase flow by *designing in the balance so that the efficacies are already in place*. **Figure 4.4** shows the impact of Lean with the increased enterprise-wide revenue per employee. This does not include the recently acquired European operations since they have just begun their Lean journey. These operations are also making great progress in their lean journey with the cross-site collaboration (As of publication, they've achieved a 28 percent reduction in LT in just six months!).

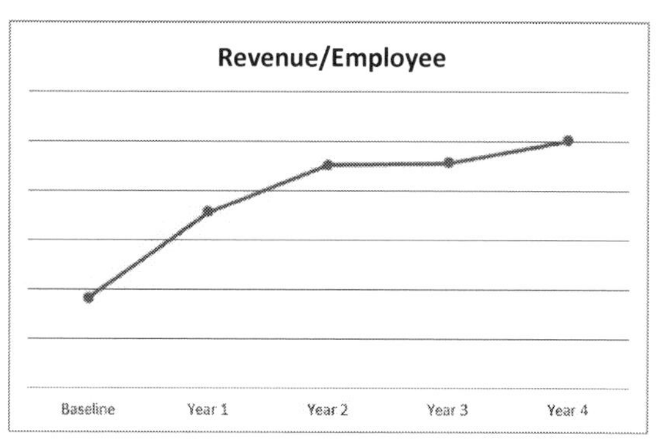

Figure 4.4

Production Map Types:

1. **Current State Value Stream Map (C-VSM):**

The C-VSM is the living depiction of the process. Shook/Harris depicts the general components of a C-VSM. There are offshoots and variations via other authors. The intent of the C-VSM is to identify the known process parameters and then to identify the "kaizens," or opportunities for improvement, to take those parameters to the next level. The very first C-VSM establishes the baseline from which the team can see its progress. Key baseline metrics might include: Lead Time, DIOH, Throughput, Productivity, Inventory Dollars (across the flow), etc. (**Figure 4.5**).

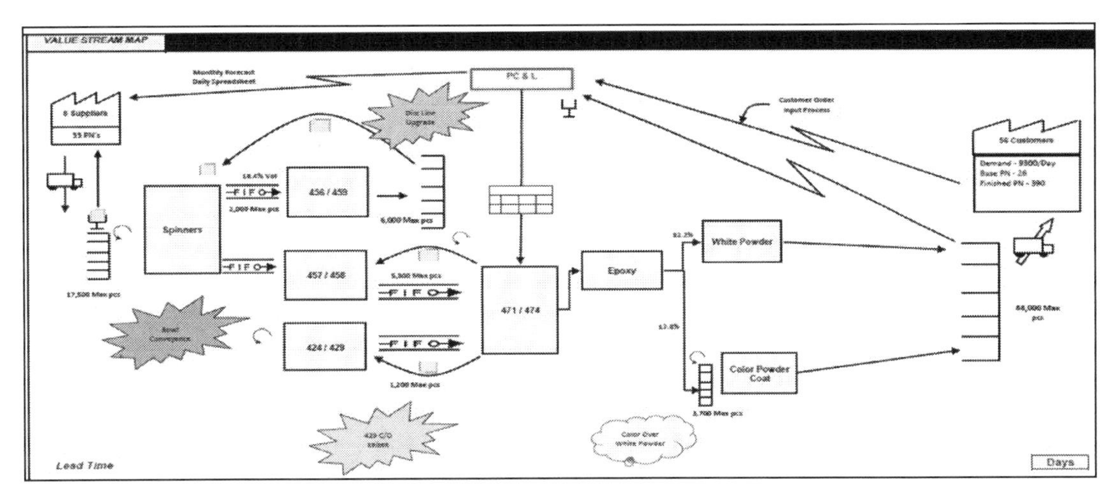

Figure 4.5

2. **Future State Value Stream Map (F-VSM):**

The F-VSM is the estimated depiction of process *sometime* in the future. Again, *Learning to See* by Shook and Rother depicts the general components of an F-VSM. The intent of the F-VSM is to show the flow of the process after the previously identified kaizens have been implemented. It may also show the future kaizens that are being planned (Capital equipment, spread of the process into other areas, etc.). **Figure 4.6** shows a sample F-VSM. Businesses have differing logic as to how far in the future an F-VSM should be assessed for. Some teams do five years, others one year. *Accuride sets their F-VSM timing for three months.* Things are so dynamic early in the journey that anything longer is not perceived as adding value in terms of the time spent in the review. Accuride's QLMS Council executes a quarterly review of <u>*every key VSM in the organization*</u>. Quarter-to-quarter comparisons are made on several KPIs: Lead Time, DIOH, Throughput, COPE and KPIV. A traveling trophy is awarded to the site that makes the biggest step-function change.

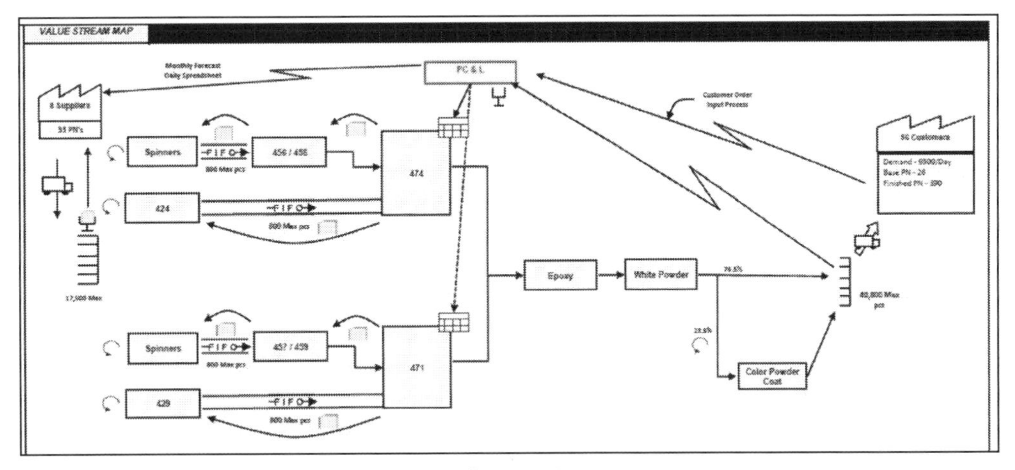

Figure 4.6

VSM's are the backbone of depicting process flow. Business leadership must be willing to accept the information provided without being demoralizing and/or applying punitive actions. They need to enable the necessary resources to support the changes identified by the team. Many personnel find that they cannot handle this level of introspective process evaluation and "opt-out" by leaving the organization. This is a win-win, which often enables the identified excess resources to back-fill into those roles.

3. **Enterprise Value Stream Map (E-VSM):**

   Discussed in more detail in Chapter 6, the E-VSM takes the business to a new level. Accuride leveraged Jonathan Chong's article "Enterprise-Wide Value Stream Mapping: Create a Vision of Your Company That Really Puts Your Customers First." After the site-based VSMs are established, the E-VSM looks at the interactions across the organization. We consider this from an "inside out" perspective. It is largely a transactional process evaluation. This rapidly enabled the business to take out another significant chunk of working capital across the business flow in a very short timeframe AND further reduced LT by another 10-20 percent. PFEPs were modified to match the Takt times and/or pull demand from the customer (**Figure 4.7**).

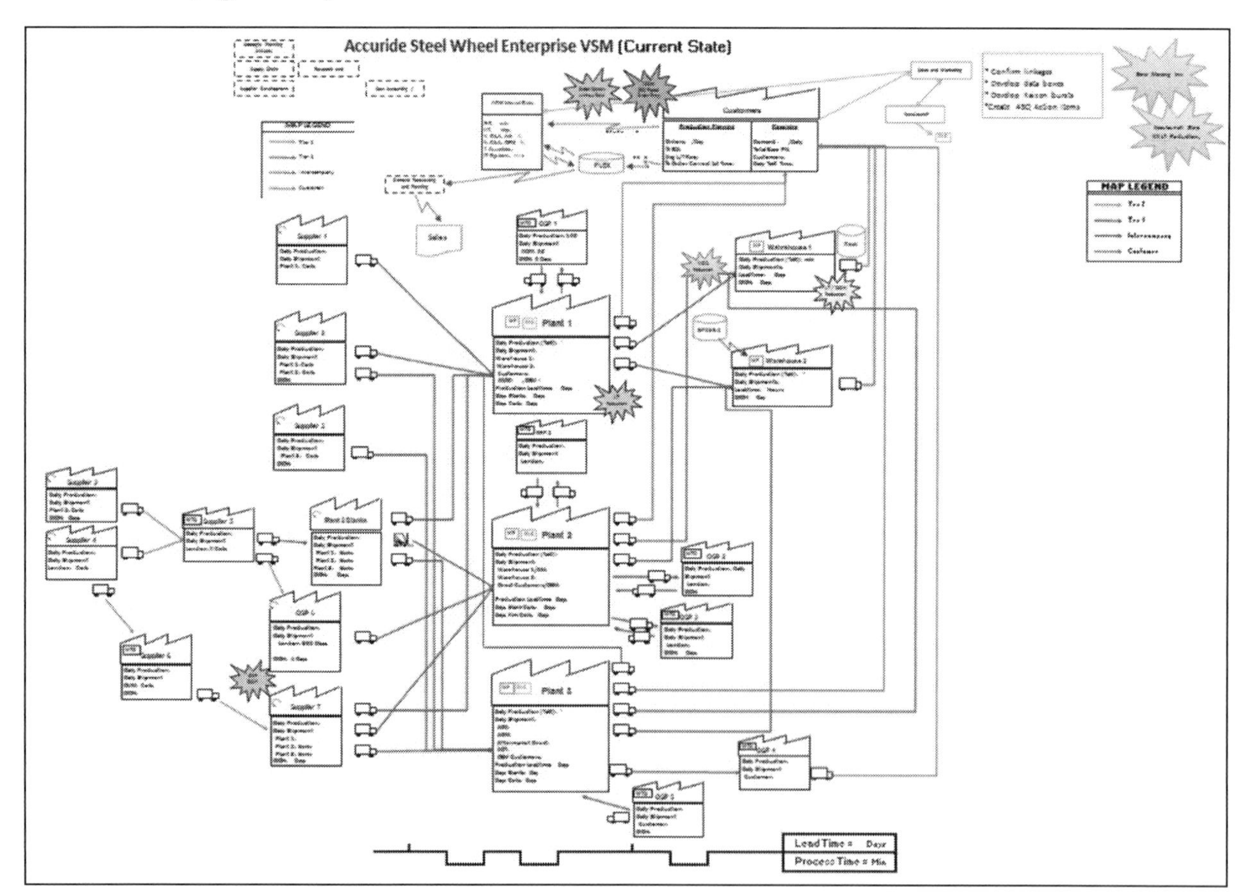

Figure 4.7

4. **Transactional Map- Non-Value Added "But Necessary" Processes:**

Transactional processes are used to depict the flow of information or data creation and delivery. They are typically human driven and have some form of trigger to start the processing of the data with an output that needs to feed another process. Information flows need to be effective (right the first time) and efficient (with the least amount of effort and when it is needed). T-VSMs can be drafted in a variety of formats that match the complexity of the process (**Figure 4.8**).

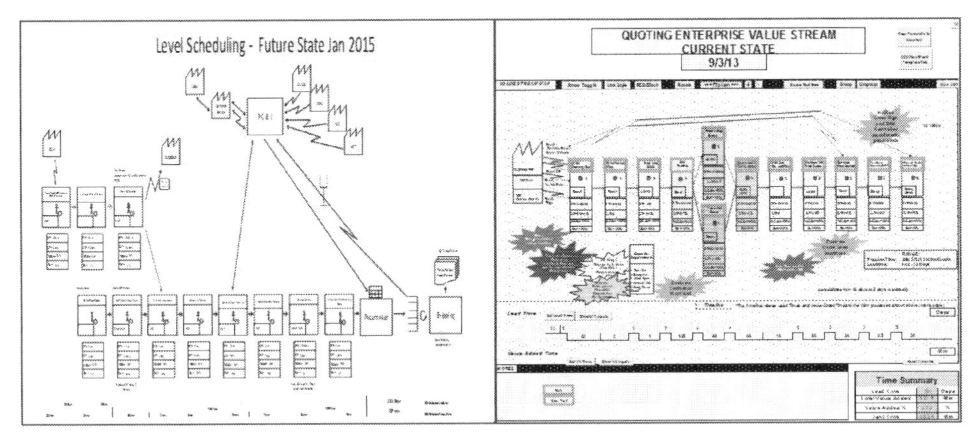

Figure 4.8

T-VSMs are conducted on core transactional value streams across the business. Each functional department exists to provide core functions. Those functions are expected to be as effective and efficient as feasible. T-VSMs have been conducted across Finance, Supply Chain, Quality Management Systems, Warranty Processing, Call Centers, Human Resources, Inter-Company Sales, Engineering, IT and others. There are functionally focused Lean books out there for these teams, as well. Lean is not just for manufacturing. Refer back to **Figure 2.14** for sample resource ideas for VSMs in non-manufacturing disciplines.

Thousands of hours are lost annually in daily transactions that are not needed if the preceeding processes are correctly performed. Management does not tolerate these types of losses in manufacturing and the "spotlight" used to illuminate these errors is like a Klieg spotlight used in the sky. However, very little is typically done to evaluate, determine root cause and implement permanent corrective actions in the transactional side of the business, causing permanent "sidecar" systems to catch and fix those issues. The key takeaway, though, is that if these processes are not optimized, then several types of losses occur:

- Production processes often depend on these outputs, which can trigger late or incorrect production and/or quantities, wrong ship-to locations, improper build methods, etc. If the transactional data is not correct or timely, then there is a limited ability to optimize manufacturing.

- Overhead costs are one thing customers are the least likely to pay for. Any opportunities for cost reduction in the transactional area literally drops to the bottom line. Think about the opportunities to reduce third party expenses in Finance, government oversight, etc.
- It's a "them" not "me" culture. This is not about Lean Manufacturing. Its about Lean Management; all of the systems need to co-exist at an optimal level. That "head" or "belly button" or "toe tag" is a person with a family and responsibilities deserving of the business' respect. The organization hired them in good faith to be a long-term part of the team.

Unlike direct labor personnel, transactional leaders often do not have layers of management over them scrutinizing every step. With that said, they must be introspective and take the initiative to ensure that their processes are as effective and efficient as possible in order to optimize the business costs in order to secure the future of the overall organization.

T-VSMs are commonly conducted via SIPOC and/or Swimlane maps. A SIPOC (Supplier, Input, Process, Output, Customer) review is typically completed at each step of the process along with key performance data. The origin of SIPOC is traced to the period around the total quality management program. **Figure 4.9** is borrowed from Google Images and shows the concept of SIPOC.

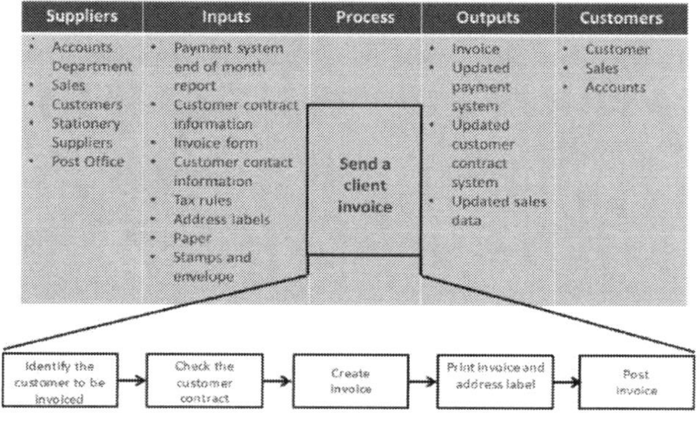

Figure 4.9

This cannot be stated firmly enough:

> <u>*Go and follow the flow*</u>. Walk, drive or fly to see each process step. The levels of distraction, error cycles, incomplete incoming information, special "yeah, but" processes, etc. are simply not effectively captured when trying to map flows when sitting in a conference room. Internal suppliers are better able to see their impact on internal customers across the full flow.

Common types of performance data could include Lead Time, Queue Time, Cycle Time, %Accuracy or Yield, %Rejects or Return requests for more information.

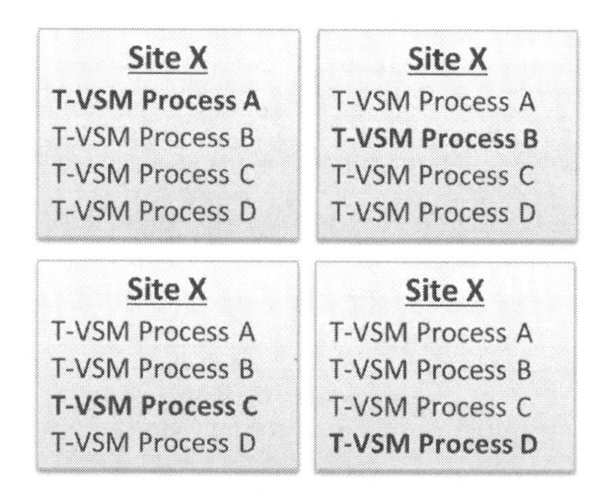

Figure 4.10

Replication. Why do it more than once?

Sites typically have similar Transactional processes in finance, HR, IT, production control, engineering, etc. As one site conducts its T-VSM on a Process, they can share that flow and lessons learned with a sister site. Accuride "dealt out" the key T-VSMs to the sites, and they each conducted two to three T-VSMs. They then shared their process flows and lessons learned to all of the other locations (In a roll out cadence). The receiving sites then were able to take the base T-VSMs and build upon those and tailor as needed to their location. Second rounds were conducted as the updated flows were completed. This prevented each site from having to conduct a full suite of T-VSMs. The reviews happened across focused process calls, Lessons Learned sessions and/or in the Quarterly Reviews (**Figure 4.10**).

Across the process, the iterative results were re-shared back into the organization often generating some form of VOS or StdW process. As the sites saw the optimizations from each other, a majority of the transactional processes became more and more standardized across Accuride without a “forced fit” approach. Alignment and buy-in happened rapidly. Most replication cycles are completed in two to three quarters.

This powerful approach enabled rapid alignment on most of the key transactional processes across the business further supporting Accuride’s velocity in lean execution.

T-VSM Mini-Series: Just a few examples

At Accuride, there have been multiple areas of self-scrutiny in the transactional areas:

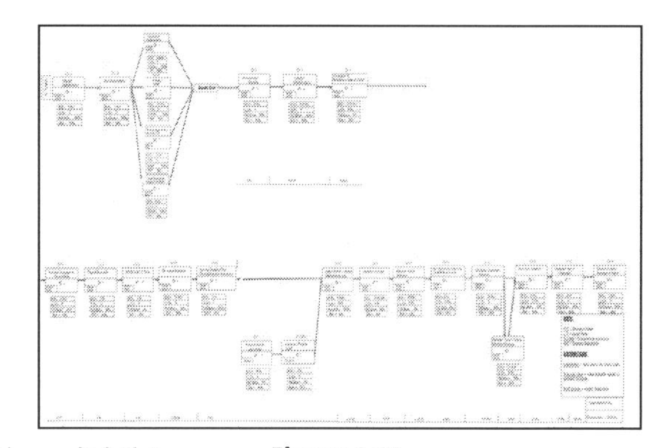

Figure 4.11

- **Financial Month End Close (MEC):** This team shortened the MEC timing process by two days while simulataneously reducing the number of internal errors generated during the process (**Figure 4.11**). Just to make it fun, they did this while simultaneously implementing a new ERP system (Plex).

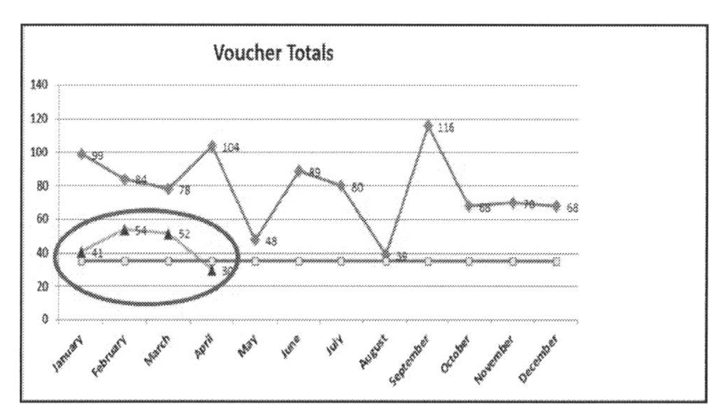

Figure 4.12

- **Financial Vouchers Issued:** This team realized that the frequency of credit voucher issuance was self-inflicted due to errors in previous steps in the process. Aside from personnel time, there were other losses incurred with these errors. The team was able to reduce the quantity of issued vouchers by more than half (**Figure 4.12**).

- **Financial Returned Goods Memo (RGA):** This team saw that it was taking months (or years) to manage customer credits via the issuance of RGAs. Customers just took the credit if these aged out. This caused a variety of other process losses, such as loss of returned products for evaluation, customer ill-will, etc. The cross-functional team reviewed flow and implemented several fixes and kanban like triggers, reducing time to issue an RGA by 65 percent (**Figure 4.13**).

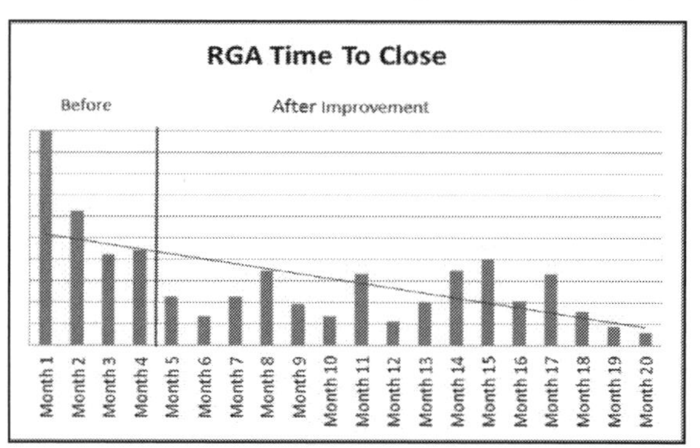

Figure 4.13

- **Engineering AccuLaunch™:** This team realized that the "vanilla" version of APQP was not optimal for product and process launches. They bolted on key process evaluation features to include Lean, EHS, Maintentance, etc. as shown previously in **Figure 1.8**.

- **Engineering New Job Start Up (NJSU):** This foundry team cut its quoting time for new jobs by more than half. NJSU started as a Design For Manufacturing (DFM) tool in a foundry setting involving the cross-functional team upon receipt of a new job. The potential job complexities were evaluated up front (**Figure 4.14**).

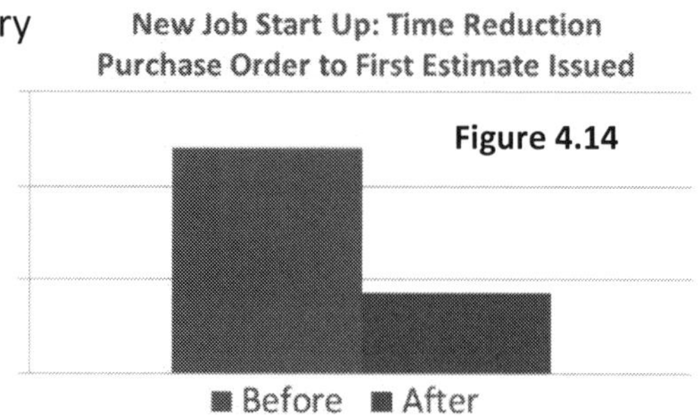

Figure 4.14

- **Warranty Management System (WMS):** Customers deserve a rapid response to submitted product concerns. The business needs a prompt response in order to manage any potential safety concerns. The team completed a C-VSM review of the WMS process. They have reduced the Median Time To Close (TTC) by more than 50 percent. Customer survey response has been positive. They analyzed the slowest process steps and used claim-to-

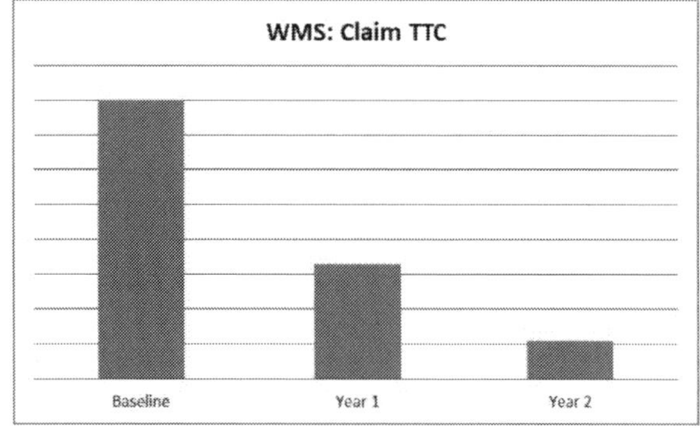

Figure 4.15

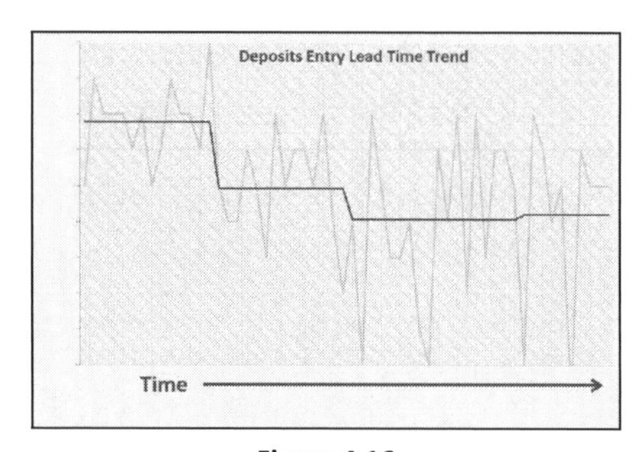

Figure 4.16

completion rates to help identify correlations between process time and special types of claims. Various customers have differing types of claim management systems (**Figure 4.15**).

- **Deposit Entry Automation:** An Accounts Receivable analyst decreased data entry lead time by 35 percent. This process was adopted across the enterprise. Redundant attachment and journal entry processes were eliminated (**Figure 4.16**).

- **Pricing Errors:** There were multiple system issues with pricing data. The team developed a new model saving about 20 percent of their time dealing with errors that were generated. A large impact was the future avoidance of debit memos.

- **Capital Asset Process:** This cross-functional team reduced the amount of time correcting the data entry process for capital asset requisitions by more than 80 percent (**Figure 4.17**).

- **Freight Optimization:** Freight was being routed manually. A new routing process was implemented significantly reducing the expense as a percent of COGS (**Figure 4.18**). In addition to reduced costs, the new process greatly reduced processing time, minimized manual data manipulation and enhanced reporting capability.

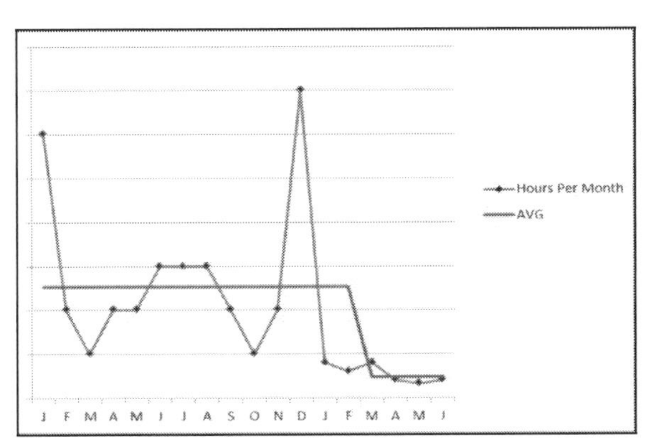

Figure 4.17

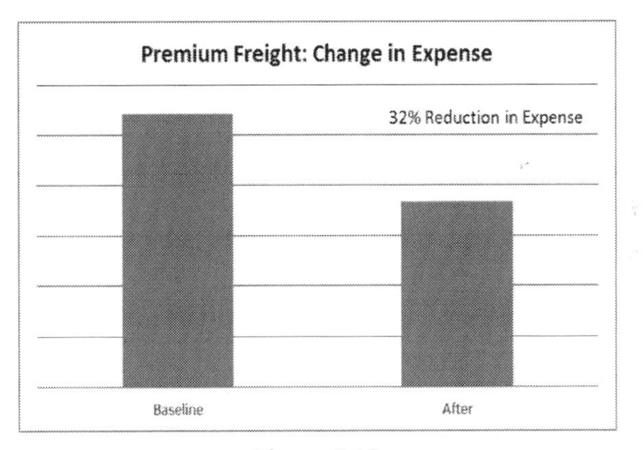

Figure 4.18

- **Foundry Overtime:** There was too much overtime spent on cleaning emissions. The team needed to organize itself and properly train the employees to perform the task. A cross-functional team provided the employees with the right tools to do the job and the waiting time was cut in half (**Figure 4.19**).

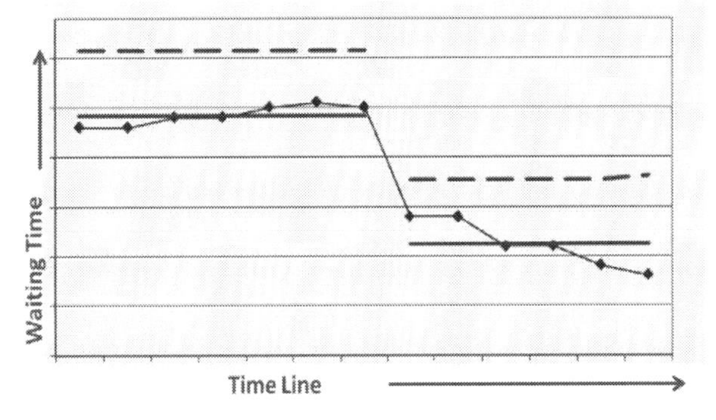

Figure 4.19

Sample Lean Tools to Support VSMs at Accuride

VSMs leverage different types of "time" in their analysis. It is important to clarify the concepts of these terms:

1. **Lead Time (LT):** The total amount of time between the recognition of a required task, operation or process and its completion. Elements can include order entry, material purchase, setup, queue, processing, shipment and other activities.
2. **Cycle Time (CT):** The total time needed to complete a transformation from one status to another. The time it takes from one finished unit to the completion of the next finished unit. One of the most frequently observed errors with CT is that people think that the CTs can be added up to equal the total LT; it almost never will. There are three key reasons: Parallel processing, Queue or Inventory Time and Rework processing. There may be others.
3. **Takt Time:** This is the pace or beat that is needed to match the process output to the customer demand. Takt is very different from CT because it is the calculation of *what is needed* where the CT is the measure of *what it is*. The goal is to balance the CT to achieve the Takt.

Finding Waste (Muda)

Lean is the relentless focus on the reduction of waste. Waste robs our business of agility and competitiveness. It is a management team's responsibility to *proactively* be able to *identify* the forms of waste *and* provide the means to *eliminate* it. No tool eliminates waste, only people can do that. There are many books out there that describe the various forms of waste. Accuride has leveraged many of those and expanded upon them using the following waste categories that fall into an acronym called **COMMIT NOW** shown in **Figure 4.20**:

- **C**orrection: Anything not done right the first time.
- **O**ver-production: Sub-Assembly. Anything made when not needed.
- **M**ess: Disorder/Chaos.
- **M**otion: Ergonomics; Having to go get things when they should be on hand.
- **I**nventory: Raw Material, WIP, FG. Product and non-product (MRO, Office, etc.).
- **T**ransportation: Wasted distance traveled.
- **N**ot Using Ideas: Lack of employee input utilization.
- **O**ver-processing: Make it better than it needs to be.
- **W**aiting: Inefficient delivery of parts or data.

Figure 4.20

Mura

Mura is unevenness: It is the main cause of waste. The herky jerkiness of production orders (**Figure 4.21**) with a variety of system losses in the mix. By smoothing out those demands via a variety of Lean tools, this effect can be mitigated. **Figure 4.22** shows how some of the Lean tools can be cross-referenced to positively impact Mura. Key approaches could include:

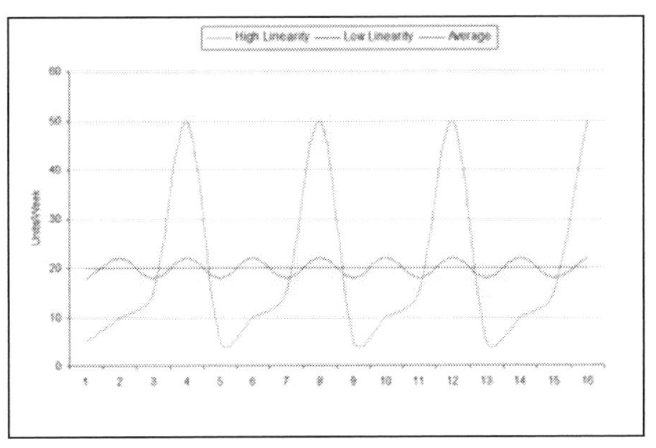

Figure 4.21

- Balancing the work flow to the Takt time.
- Establishing the PFEP/TTR to plan the markets.
- Effectively executing Set-Up-Reduction.

Muri

Muri is overburden: Piling resources up with unreasonable demand can lead to multiple forms of losses from safety issues to scrap to equipment breakdown. Various forms of StdW (Machine/Operator/Leader/IDL/OH) can be conducted to ensure that the system has the necessary levels of human interaction to be effective... and efficient (**Figure 4.22**). Key approaches could include:

- All forms of StdW evaluation.
- Equipment: Total Productive Maintenance, Set-Up Reduction, Autonomous Maintenance, etc.
- Jidoka.
- Tugger/Material Delivery Systems.

Figure 4.22 Summarizes some common Lean tools utilized at Accuride with respect to how they impact Muda, Mura and Muri. There are more tools than these, and this summary acts as a VOS reference guide for the Accuride team.

General QLMS Process Tool Impact Guidance Table	Muda	Mura	Muri
Process Tool	Waste	Uneveness	Overburden
5S	X		X
8D CA/PA: Corrective/Preventive Action	X	X	X
A3 Problem Solving	X		X
AAA: Attribute Agreement Analysis	X		
A-TPM: Autonomous TPM		X	
C-VSM: Current Value Stream Map	X	X	
E-VSM: Enterprise VSM	X	X	
F-VSM: Future VSM (3 mos)	X	X	
Kan Ban; Continuous Flow		X	X
KPIV: Key Process Indicator Variable	X		X
MBal: Machine Balance		X	X
OpBal: Operator Balance		X	X
Pace Maker		X	X
Paynter Charts: Predictive Warranty	X		
PFEP: Plan For Every Part	X	X	X
Plant/Cell Layout	X	X	X
Replication: Similar Processes	X	X	X
Right-Sized Markets	X	X	X
Southwest Rule Addressing System	X	X	
SPC: Statistical Process Control	X		X
StdW: Standard Work	X		X
TPM: Total Productive Maintenance	X	X	
TTR: Time To Replenish	X	X	
Tugger	X		X

Figure 4.22

Setup Reduction (SUR) or Rapid Change Over (RCO)

When a line is performing one type of work and needs to perform another type, a change of setup is involved, which results in downtime for the line. This downtime is a form of lost capacity, and all efforts are required to minimize it. Set-up time is the elapsed time of when the last good piece is run and the first acceptable new piece is complete.

The "order" in which SUR/RCO can be performed across a day or week of production can be optimized from "easy to hard" or vice versa to simplify the transition from one product type to the next; especially if the products are "family" based. Coordinating this potential with the PFEP/TTR process can be quite beneficial. Some key SUR/RCO benefits include:

1. Faster response to customer demand: Changeover becomes transparent to the process flow... More ability to run level loading.
2. Improved Quality: No "batches" of bad parts... Fixes are made immediately.
3. Less inventory required: WIP between operations is reduced/eliminated... Lower levels of FGs are required.
4. Reduced Operating Costs: Minimal non-value add activity... Increased productivity and capacity, better tool control, minimal cost to execute.

Poka Yoke/Error Proofing

Poka yoke, or error proofing, is the application of tools to **prevent** an error from occurring. Several levels from detection to prevention can be stepped through in order to attain the ultimate level of Poka Yoke, which is prevention. **Figure 4.23** shows a simple train crossing sign, which acts as a visual cue. The second sample shows a mechanical gate, which, while more effective, doesn't prevent someone from trying to do an end run. However, the true level of Poka Yoke is achieved at the third sample where the train is on a trestle and cannot meet the traffic. As teams work to build quality into the process as the "process experts"—no, not the engineers, but the people that work with the process all day long—ask them what they would do so that "x" could never happen.

Figure 4.23

These other tools can be used to facilitate error proofing: Checklists, Templates, Photos, Mechanical Stops, Video Instructions, Color/Shape Coding and StdW.

5S

5S Focuses on improving the immediate environment by removing mess and minimizing lost time by waiting and searching. A clean environment also improves morale (**Figure 4.24**).

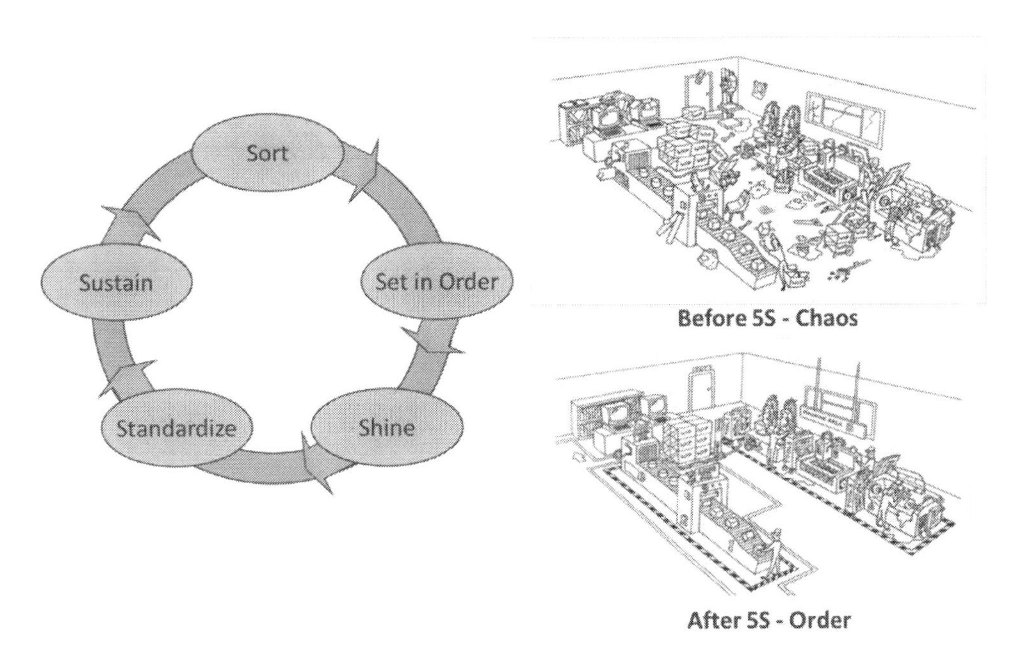

Figure 4.24

Key Takeaways

1. The backbone for getting effective flow is to understand every step in the process.
2. There are generally two process types: Transactional and Operational. Both need to be performed effectively (right the first time) and efficiently (with minimal waste).
3. Two VSMs are generally prepared: current state with opportunities for improvement via kaizen bursts and the future state (three months out) when we expect certain kaizen burst actions to be in place.
4. An Enterprise Value Stream Map ties the impact of all the individual VSMs for the plants and other supporting processes together, discussed more in depth in Chapter 6.
5. SIPOC is a very powerful tool that ties the process steps to its output and customer at one end and the inputs and the sources of these inputs at the other.
6. Transactional VSM examples include month end close, return goods authorization, new product development, new job start up, warranty time to close, deposits entry, capital asset process and foundry overtime reduction.
7. A number of Lean tools can be used to make improvements to the value streams.

Chapter Five: Visual Operating Systems (VOS)

Visual Operating Systems (VOS) are the backbone of an effective Lean system. They are, by name, visual in nature and are meant to enable an "at a glance" understanding of situational status. VOS are an efficient method of information sharing with teams, leadership and visitors. The data is expected to be self-explanatory, and the need to ask (resulting in process disruption) is minimized. Facility or process leadership looks at the VOS and is immediately able to ascertain whether or not additional support or evaluation is needed. General VOS impacts include:

- Live operator awareness of process status. Issues are apparent for immediate resolution.
- Visually objective process status to any team member reviewing the process.
- Non-confrontational sharing of data: Facing a VOS board side by side is considered to be a more effective communication style when sharing potentially negative data face to face (as the focus moves from individual to the data).

Accuride's VOS tools are roughly sorted into two categories:

1. **Corporate-wide Standard:** These are required forms of StdW that have been sourced internally or externally and adopted by Accuride's QLMS Council. All sites must use and apply these tools in a prescribed methodology. Any manager from any Accuride location can immediately identify with the VOS and understand its inherent message. These core forms of StdW are integrated into Accuride's culture.
2. **Optional VOS:** These forms of VOS do have a StdW format as to how they are supposed to function. However, due to the nature of the facility and/or volume of flow, the immediate look and feel of these tools can vary so long as they effectively communicate their intent.

This chapter will provide a high level overview of some of the key forms of VOS that are actively leveraged across Accuride. We don't consider these forms of VOS to be new or unusual in the world of Lean. These are some of the ones that have worked very well for us in our sites and within our culture as a result of training and continued focus. They are culled from a variety of sources: Experience, Consultants, Benchmarking Visits, Conferences, Books, Lean Websites, etc.

Accuride's QLMS Council reviews these forms of StdW annually to determine if upgrades are needed, if new forms of VOS are learned for potential modeling or if some forms of VOS can/should be retired. Each VOS has a written procedure on its general use and intent. These procedures are part of Accuride's QLMS as a part of the Improvement section within ISO 9001:2015 and IATF 16949:2016.

There are dozens, if not hundreds, of potential forms of VOS. The ones highlighted in this chapter are presented to provide a visual snapshot of what these tools look like and how they work at Accuride.

The "Glass Wall"

What It Is: A high-level snapshot of the facility at a monthly level is shown in **Figure 5.1**. KPI performances regarding Safety and QCD are shared. The C-VSM for the site and its F-VSM are posted. Current kaizen plans are shown on the VSMs as well as key action plans to accomplish the Future State. F-VSMs reflect the changes anticipated in the near future (the next three months) and not too far so that we can evaluate the effectiveness of the actions and correct them quickly if we do not achieve the expected results.

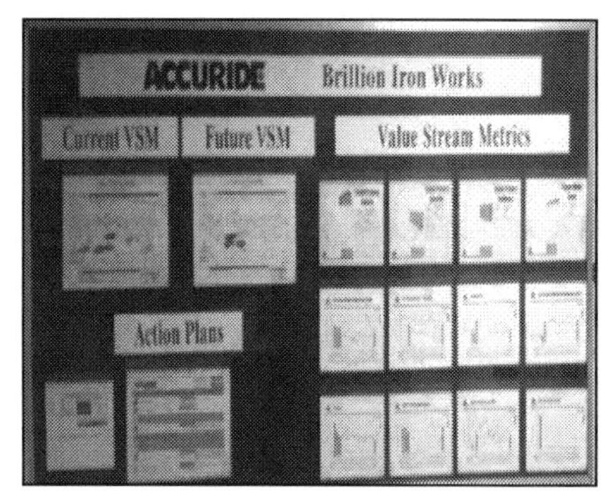

Figure 5.1

Corporate or Optional: This VOS is a Corporate form of StdW that every Accuride facility is required to have in place. It is one of the first places that visiting leadership stops at to get a sense of overall facility performance.

Who Uses It: Key metrics are shared with all associates, visitors and customers in a prominent area.

Value Stream Maps (VSMs)

(See Chapter 4 for more details and sample insets)

What It Is: An in-depth analysis of a process flow. Accuride leverages several forms of VSMs depending upon process type and scope. These are previously described in Chapter 4. There are four general types of VSMs that are utilized at Accuride:

1. Production-based C-VSM.
2. Production-based F-VSM. Three months after the C-VSM.
3. Transactional process-based maps. T-VSMs are intended to map data and/or information flow. It can take many forms.
4. Corporate-wide product-based E-VSM.

Corporate or Optional: Manufacturing-based VSMs, both current and future states, are drafted in a required StdW Excel format. T-VSMs are optional and follow a variety of methods depending upon process complexity. E-VSMs also follow a standard template integrating key elements of the operations across the product category, across multiple facilities and functions.

Who Uses It: VSMs are developed by the process owners and reviewed regularly. Accuride conducts a quarterly Lean review where all key VSMs are shared across the business.

Plan For Every Part (PFEP, *Making Materials Flow*)

What It Is: PFEP is a key material planning tool to optimize the amount of inventory needed at every stage in the process flow. It is a linear flow of data in one location to capture how each part is purchased, received, packaged, stored and delivered to its point of use. While this data generally exists in an organization, it is usually stored in many different places under the control of different personnel. The PFEP is the first step in creating a lean material-handling system for purchased parts. ***It results in a rate of units required per unit time (usually to the nearest hour).*** **Figure 5.2** shows sample table headings for a PFEP while **Figure 5.3** shows a model.

(Some **Figure 5.3** columns are deliberately obscured)

Part Number	Container Weight	Standard Container Quantity
Description	Single Part Weight	Shipment Size
Daily Usage	Container Dimensions (LxWxH)	Carrier
Storage Location	Qty per Assembly	Transit Time
Order Frequency	Hourly Usage	Supplier Performance
Supplier	Containers/Hour	Kan Ban Cards in Loop

Figure 5.2 (Source: *Making Materials Flow*, Harris, Harris, Wilson)

Supplier to Purchased Supermarket															
P/N	Avg Daily	Shipment Size	Largest Pull During Shipment Size	Internal Variation	External Variation	Std Pack Qty	Max	Max Pieces	Transit Time (Day)	MIN	Supplier	Std Pack Qty Width	Std Pack Qty Length	Std Pack Qty Height	Market Location
50408PKBLK21	165	1	308	143	165	24	20	480	0.5	83	Accuride Henderson	48	48	60	
[illegible]KWHT21	8	[illegible]	[illegible]	[illegible]	[illegible]	24	10	240	0.5	16	Accuride Henderson	48	48	60	
[illegible]PKBLK21	101	4	618	214	101	24	30	720	0.5	51	Accuride Henderson	48	48	60	
29001PKGRY21	2	4	18	10	2	24	1	24	0.5	1	Accuride Henderson	48	48	60	
[illegible]WHT21	1	[illegible]	[illegible]	18	1	24	1	24	0.5	1	Accuride Henderson	48	48	60	
[illegible]W	8	[illegible]	[illegible]	23	8	24	3	72	0.5	4	Accuride Henderson	48	48	60	
[illegible]PKBLK21	1	4	4	0	1	1	5	5	0.5	1	Accuride Henderson	48	48	18	
[illegible]	[illegible]	[illegible]	4	0	1	1	5	5	0.5	1	Accuride Henderson	48	48	18	
[illegible]	[illegible]	[illegible]	10	2	2	1	12	12	0.6	1	Accuride Henderson	48	48	18	
[illegible]PKWHT21	1	8	24	16	1	1	25	25	0.5	4	Accuride Henderson	48	48	18	
[illegible]PW	35	[illegible]	[illegible]	820	35	24	48	1152	0.5	140	Accuride Henderson	48	48	60	
[illegible]W	1	[illegible]	[illegible]	8	1	1	13	13	0.5	2	Accuride Henderson	48	48	18	
[illegible]PKBLK21	9	4	224	198	9	1	243	243	0.5	18	Accuride Henderson	48	48	18	
50408PG	2	1	28	26	2	1	30	30	0.5	1	Accuride Henderson	48	48	18	
50408PKWHT21	34	[illegible]	[illegible]	[illegible]	34	24	5	120	0.5	17	Accuride Henderson	48	48	60	
50408PW	237	1	499	262	237	24	31	744	0.5	119	Accuride Henderson	48	48	60	
50408PKYEL21	9	1	48	39	9	1	57	57	0.5	5	Accuride Henderson	48	48	18	
[illegible]PKBLK21	5	[illegible]	[illegible]	[illegible]	5	1	35	35	0.5	3	Accuride Henderson	48	48	60	
[illegible]PKGRY21	50	1	281	231	50	24	14	336	0.5	25	Accuride Henderson	48	48	60	
50487PKWHT21	52	1	204	152	52	24	11	264	0.5	26	Accuride Henderson	48	48	60	
[illegible]	[illegible]	[illegible]	[illegible]	[illegible]	[illegible]	24	3	72	0.5	5	Accuride Henderson	48	48	60	
50408PKGRY21	27	1	228	201	27	24	11	264	0.5	14	Accuride Henderson	48	48	60	

Figure 5.3

The PFEP is the backbone upon which material handling, planning and site flow are established. Data columns in a PFEP may vary from business type to business type.

Corporate or Optional: A PFEP is a required VOS in Accuride. PFEPs are established for all parts, including raw materials, WIP, FG, Tooling, Dunnage and MRO supplies. PFEPs are reviewed on a regular basis.

Who Uses It: PFEPs are developed by the PFEP Manager and are reviewed regularly. They are a critical input to the AccuLaunch™ process when determining material flow into, throughout and out of the facility.

Time To Replenish (TTR, *Making Materials Flow*)

Figure 5.4

Supplier	Item	MaxPull	4 Weeks Sales	Daily Usage/20	1/05 thru 2/02	(Pcs)Daily Usage	TTR Shipment Size(days)Modified EPEI	BPCS Lead Time	Standard Pack QTY	Largest Pull During Modified EPEI	Internal Buffer - Demand Variation (pcs)	External Buffer	Calculated Max Inventory (Pcs)	80% Target
	Some columns deliberately obscured						10	21	24	44	0		44	35.2
							10	21	24	194	0		194	155.2
							10	21	24	20	0		20	16

What It Is: A TTR is the frequency of when a part should be produced based on the PFEP. It is *adjusted for the variability of demand* and enables stratification between "Runners, Repeaters and Strangers." In production control jargon these might be called A, B and C items. Strangers are often "Make To Order" (MTO) products. It has other names such as Heijunka or Level Loading (**Figure 5.4**). TTR is often co-managed with the maintenance team to optimize changeover efficiencies.

When running Runners, Repeaters and Strangers across the same set of assets, time allocation is key... how many of each type are needed before changing over to the next one? TTR is a refined and calculated ballet. By optimizing changeover time and determining what the real market size needs to be, TTR is a very strong weapon in the Lean arsenal. The TTR is optimized by real world factors such as supplier reliability, delivery reliability, customer pull history, OEE, Changeover time and other features. By combining this mix of factors, an Every Part Every Interval (EPEI) is determined, which is the time interval needed to run through every offering in the product family.

The EPEI is used to determine the number of kanban cards and/or pull signals that will be used to trigger a pull from the previous process. Usually a schedule "wheel" or "flat wheel" is used to visually convey the TTR results.

Corporate or Optional: TTR is a required VOS in Accuride. TTRs are established for all parts from raw material components to WIP to FG. Dunnage, MRO supplies and tooling are also included on TTRs. TTRs are reviewed on a regular basis.

Who Uses It: TTRs are developed by Production Control in conjunction with the PFEP Manager. They are reviewed regularly as input factors change. They are a critical input to the AccuLaunch™ process when determining material handling needs, ergonomics, changeover criteria, autonomous maintenance actions and more.

Changeover Board: (Schedule Wheel or Flat Wheel or Heijunka Board)

What It Is: A changeover board is used as a visual indicator to provide the production team information as to *what part they are making now and what part to produce next*. It also provides an indicator of inventory "health" based on the market needs as calculated by the TTR. The boards are populated with kanban cards (See later in this Chapter). The quantity of cards are determined during the TTR analysis (**Figure 5.5**). The picture on the right is an *electronic kanban board* established within Accuride's Plex ERP system, which is visible to the team conducting the "pulls".

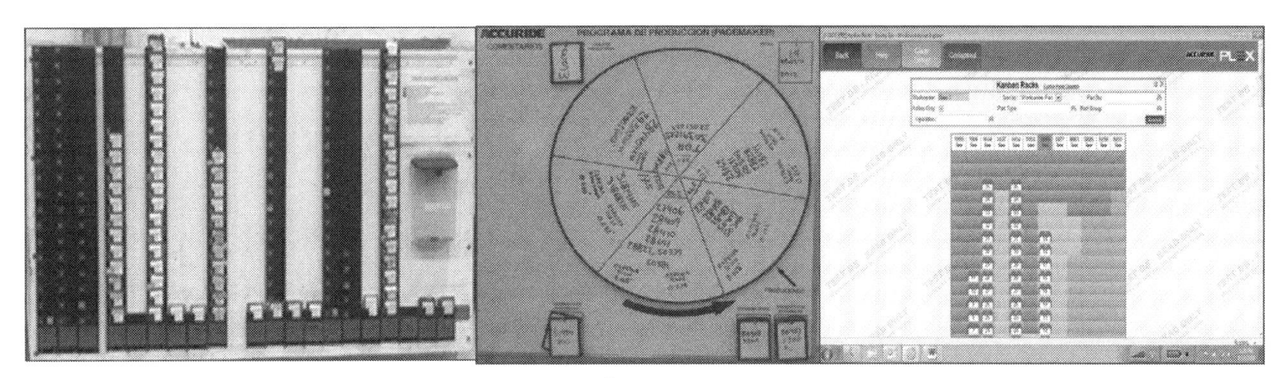

Figure 5.5

There are generally three types of kanban cards: FG, WIP or Spike (Manual). Any or all may be used to reflect production needs on a changeover board. A card basket repository is needed for the kanban cards after pull and delivery actions are completed. The baskets are emptied by specified personnel on a planned frequency. Each card is critical. If one is lost, the planned market size and overall process flow can be negatively impacted.

Changeover boards have color-coded or numeric markers to trigger the change to the next product and/or to execute an equipment changeover. Many synergies can be developed here with the maintenance and/or process changeover team. Changeovers can evolve from hard to medium to easy or vice versa to minimize the changeover time and optimize process flow. Optimizing the use of "soft" changeovers vs. a "hard" changeover across the TTR cycle can save hundreds of manhours while simulataneously enabling production agility.

Tip: Design kanban cards to be larger than "pocket" size. This reduces the opportunity for them to get inadvertantly carried away.

Lesson Learned: It is very easy to get pulled (no pun intended) into automated kanban. The Catch-22 is that the instant visibility is lost. There is tremendous intrinsic value in the touch, feel and carry of kanban cards as demand is pulled, product is produced and it is then delivered. There is certain ERP software (such as Plex) that support visual look and feel of a live visual kanban board. Some Accuride teams are modeling this to see if they can sustain the same effectiveness that has been achieved thus far.

Corporate or Optional: Changeover boards are a required VOS in Accuride. They can be optional in their design while having the same functionality. This concept enables us to rapidly assimilate new sites into our Lean systems. The boards have been modeled off of dozens of external examples and then optimized to meet Accuride's needs.

Who Uses It: Schedule boards are used by production and production control teams. The boards enable in-situ awareness as to which product is to be produced next. If by some chance the operator runs out of work, then they are assigned to other tasks to support another team.

Kanban Cards

What It Is: A kanban card is a signaling device that gives authorization and instructions for the team to withdraw items in a pull system. Accuride kanban cards (**Figure 5.6**) trigger production actions based on identified market needs (The triangular card is for batch processes). The cards stay with the product and are used to identify it in its current state until it is moved to the next location. When the product is "used" (pulled to next locale and/or shipped), the card is returned to its designated location to trigger a pull of more of the same. The poetry of this process is that *there is simply no overproduction*. Production can only produce to the quantity of the cards and then no more. Spike cards can be used to create bridge inventory when special circumstances are warranted, such as Special Sales, Holidays, Maintenance shut-downs, etc. Working capital tied up in Raw Material, WIP and FG is slashed in quick order when kanban cards are effectively established. **Figure 3.12** shows the reduced inventory trend that Accuride has enjoyed largely due to this tool.

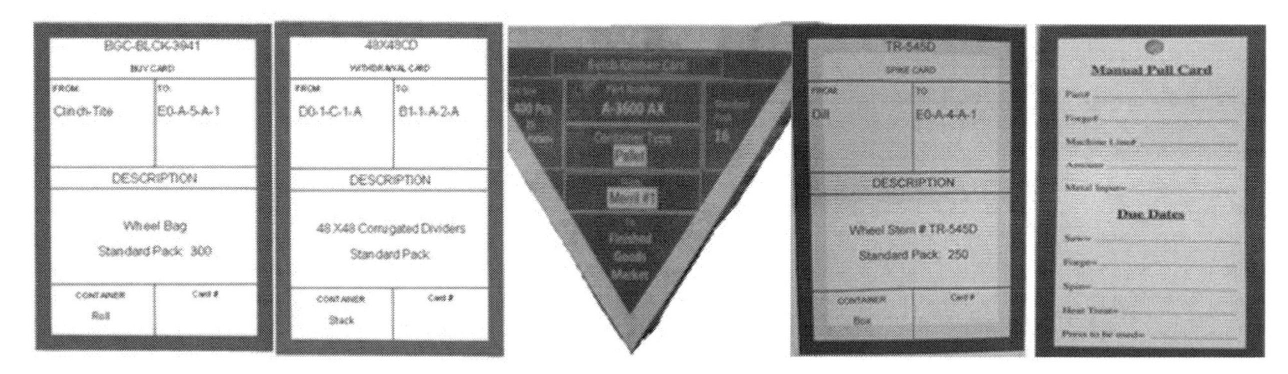

Figure 5.6

Handling logistics are dramatically reduced. Several of Accuride's sites have been able to reduce their rolling stock (forklifts) by 30-50 percent.

There are generally four types of kanban cards:

1. **FG:** Used to pull finished goods based on customer demand and markets established via the PFEP and TTR.
2. **WIP:** Used to pull sub-assemblies, components or raw materials from their markets. These markets are established via the PFEP and TTR. After process maturation, these cards are sometimes replaced with First In First Out (FIFO) lanes.
3. **Spike and/or Manual Pull:** Used as a special cause insert into the process to adjust the market based on interim information. These can be used for creating bridge inventory, managing a special sales events, supporting an unusual customer order, etc.
4. **Batch:** These are versions of FG and WIP kanban cards and are used to handle large quantity processes such as molding, forging, chemical processing and painting operations.

Remember, a card basket repository is needed for the kanban cards after pull and delivery actions are completed. The baskets are emptied by specified personnel on a planned frequency. Each card is critical. If one is lost, the planned market size and overall process flow can be negatively impacted.

Corporate or Optional: Kanban cards are a required VOS in Accuride. The cards have a similar look and feel across the corporation. This enables us to rapidly assimilate new sites into our Lean systems. The kanban cards have been modeled off of dozens of external examples and then adjusted to meet Accuride's needs. Kanban cards are heavily audited for quantity and condition. They are replaced as needed.

Who Uses It: Kanban cards are developed by the Production Control team in conjunction with the PFEP Manager. They are reviewed regularly for how many are needed as the PFEP and TTR change. Cards can be added or pulled based on the determined market sizes. Accuride has had several samples of co-developing PFEPs, TTRs and resulting kanban cards or Fax-Bans with both suppliers and customers. In almost every case overall inventories have been reduced while turns have increased.

Cell Kiosks

What It Is: A Cell Kiosk is a team level communications board with a minimum of five areas of required information sharing: Safety, Quality, Delivery, Cost and Communication. This data is reported daily for the team to see. Monthly trends and countermeasures are identified if items are not meeting the target. While the graphics themselves are the same across Accuride, the board format can be flexible depending on the work area as shown in **Figure 5.7**.

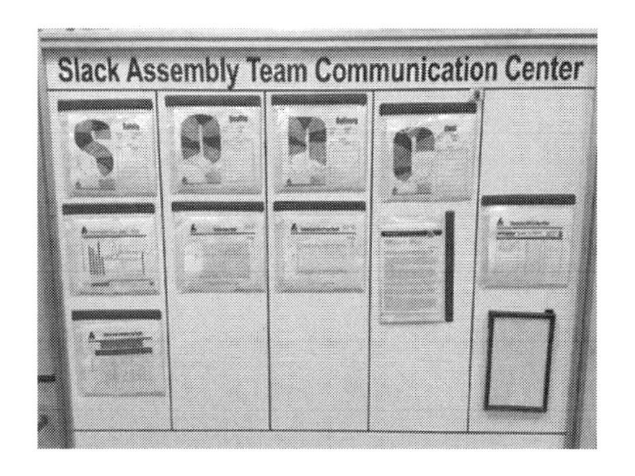

Figure 5.7

Corporate or Optional: Kiosks are a required VOS in Accuride. They are optional in their design but have a minimum report out of five components. Additional information is up to the site and/or the local team. Initial Kiosks were required to be four-sided, rotational units. These were found to be mechanically fragile and difficult to clean around from a 5S perspective. "Flat Kiosks" are more practical.

Who Uses It: Team Kiosks were developed by Accuride's QLMS Council and are utilized daily by the individual cell team and its facility leadership. Kiosks are updated daily for performance and overall performance trends of processes within the span of control of the team are shared.

Hourly Board

What It Is: An Hour by Hour Board is used to actively record process and yield information. Most often used in production, the board also has uses in repetitive transactional areas. The board has standard work sections for completion and is usually recorded in a color code format.

Across businesses, these boards can vary greatly, Accuride's Hourly Boards, sample shown in **Figure 5.8**, manage the following content:

- The hour or time of process results (usally at the top of the hour).
- Theoretical Yield: The potential maximum based on equipment specifications.
- Target Yield: Planned result; Often adapted for OEE, planned site actions, etc.
- Actual Yield: Results for that timing. **Red ink** when Target is not achieved, **Green ink** for when it is. As the hours elapse, the data is accumulated.
- Reject/Scrap: The actual number of defective units (not right first time – even if recovered). Zero findings recorded in **Green ink**, anything above zero to be reported in **Red ink.**
- Comments Section: Completed by operators, set-up personnel, maintenance, team leaders, etc. to provide any detailed information as to why targets were not achieved.
- Action Plans/Countermeasures Section: Responses to the items recorded on the board.

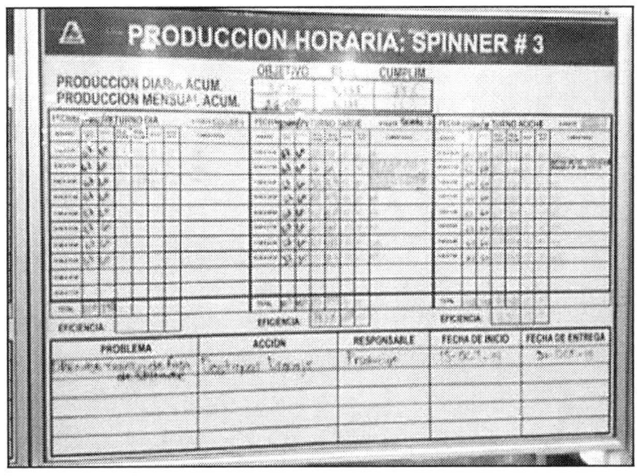

Figure 5.8

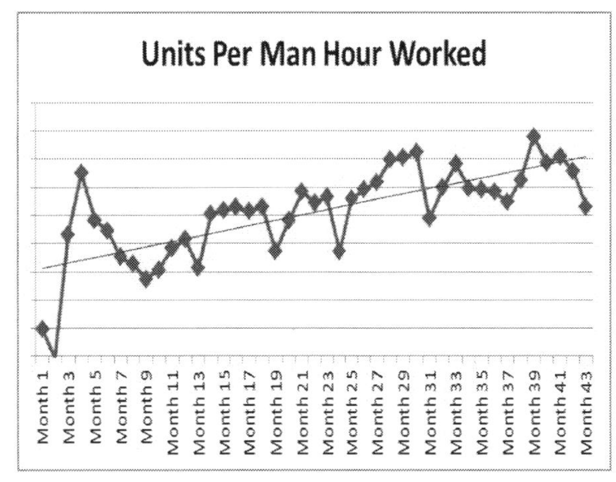

Figure 5.9

Corporate or Optional: Hour by Hour boards are a required VOS in Accuride. These Boards have the same look and feel across the corporation. Anyone from any facility can rapidly evaluate a team's process status at a glance. This powerful tool enables live awareness for the team and proactively identifies when things are or are not going well. It forces an action when needed.

Who Uses It: The local team and immediate area supervisor controls the information flow on the Hour by Hour board. In all areas, Hour by Hour boards have increased productivity when accompanied by an engaged facilities leadership team (**Figure 5.9**).

Total Preventive Maintenance (PM) Board

What It Is: The PM Board, examples shown in **Figure 5.10**, is used to clearly describe the site-based maintenance activity that is needed to support safe and effective process operation. **Figure 5.11** shows the general MTBF results since baseline. All key assets are listed on the board, and a series of colored magnets are used to identify the PM status of those assets. Operators can see when their equipment is scheduled and provide supportive advise.

- Yellow magnets indicate what units are scheduled when.
- Green magnets indicate the on-time completion of the task.
- Red magnets with green indicate that the task was completed late.
- Black magnets indicate assets with the most unplanned downtime. Teams then review the overall causals for preventive action and countermeasures.

An important measure of the effectiveness of this process is MTBF and on-time PM completion percentage. Other key KPIs may include Mean Time To Repair (MTTR) and "%Fire-Fighting vs. Planned."

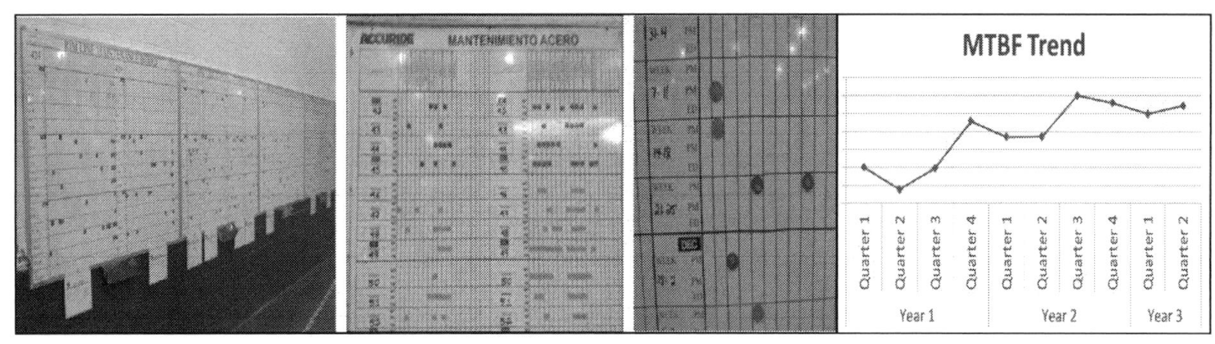

Figure 5.10 Figure 5.11

Corporate or Optional: PM Boards are a required VOS in every Accuride facility with key production assets. The boards have the same look and feel across the corporation. These boards are home grown at Accuride, and we've not yet seen them used at other businesses.

Who Uses It: The PM Boards are used by maintenance personnel and reviewed by the hourly personnel that run that equipment. The Boards are reviewed regularly to ascertain which assets need what type of PM and when.

Color Code Scheme

What It Is: Accuride leverages a standardized color scheme, shown in **Figure 5.12** for its equipment and facility structures for general visual communication across its sites. This has been a gradual adaptation. As each area underwent some type of focused improvement activity, it was upgraded to meet the standard. Over the past five years, a majority of the site work areas have been upgraded.

Green	Productive Materials and Safety	Any equipment,co ntainers, totes, carts and/or flow racks used for raw, wip, or finished materials, unless otherwise specified by customer. Safety equipment, first aid stations, and recycling containers.
Blue	Equipment/Non-Productive Material	Trash cans & the locations, tool carts & the locations, glove bin locations, fans, board locations (ie; pull, production, shipping boards) and inventory racks. Steel wheel equipment already re-painted, Iron, Foundry, Forge and Casting equipment.
Red	Non-Conforming Material and Fire Equipment	Scrap bins and locations, scrap carts and locations, salvage crib location, chip hoppers and locations. Fire fighting equipment and locations, sprinkler piping, emergency stop buttons.
Yellow	Aisleway Striping and Caution	Aisle lines - pedestrian and fork lift traffic, footprints (pedestrian path), hashed areas (no forklift traffic), black and yellow stripe - hazardous materials/chemicals. Machine guarding, handrails, guardrails, cautions and hazards (tripping, falling and striking).
Orange	Rework/Hold for Disposition and Caution/Warning	Areas for machines that need repair. Teardown carts, containers or areas. Areas for in-process product that are 'Hold for Rework', or that need disposition to determine good, rework or scrap. Re-work carts, containers or areas. Caution / Warning on machines or equipment taht could cause cutting or crushing injuries.
White	Equipment and Housekeeping	Gage carts, fixtures, communication, pull, production, tool and shadow boards. Stairways, directions, housekeeping markings, trash can markings. Aluminum forge and Iron machining, stamping, bending and forming equipment. NOTE: Steel wheel equipment will transition to white as it becomes in need of re-furbishing.
Gray	Die Racks	Die racks.
Transparent	Floor Base	All floors, including aisles, production areas, material markets, maintenance areas, unless the concrete is in unacceptable condition which would dictate another type of re-conditioner or sealer.

Figure 5.12

Corporate or Optional: The color code scheme is a required VOS with accommodations for local country codes. This enables our sites to have the same look and feel across the corporation.

Who Uses It: The Maintenance, Operations and EHS teams largely manage and implement the Color Scheme. As equipment is moved, refurbished or systems are upgraded, the color schemes are executed. This a part of the overall 5S system within Accuride.

5S Audit/Footprint Board

What It Is: 5S is not housekeeping. Many organizations start with 5S as their first step in Lean. In Accuride, we first started with VSM. We found 5S to be a wonderful "side effect" of executing Lean. Those first broad-sweeping 5S efforts can uncover a lot of unused materials for re-sale and/or internal use generating cash for self-funding opportunities. We've seen sites "e-Bay" antique gauges and office furniture, sell scrap metal, recover years of office supplies, find salable product, expensive gauging and tooling, etc.

As VSMs are executed and process flows are laid out, PEEP is a natural by-product (a Place for Everything and Everything in its Place). Many consider the StdW of cleaning up as non-value added. That's technically true. However, it is also a strong form of cost avoidance. Mess makes people have to look for things, which robs the process of the operator's time. Mess loses items and triggers re-orders and shrink. Mess creates an unsafe work environment causing accidents and spills. Mess lowers employee morale.

A 5S board is essentially a footprint of the facility. It includes a plan view of the site location showing all structures and grounds. There is some form of pass/fail color coding of internal audit results. Audits are conducted on a regular frequency by all levels of management to ensure fresh eyes and process sustainment.

5S Audit Boards (**Figure 5.13**) are often combined with Safety audits and Quality System audits. Performance data stays on the board for a rolling 12 months. A cool adaptation in the most the most recent months is the ability to do audits via an audit scan code (**Figure 5.14**). The auditor can scan their code, pull up an automated evaluation form, complete the assessment on their phone and then return the data.

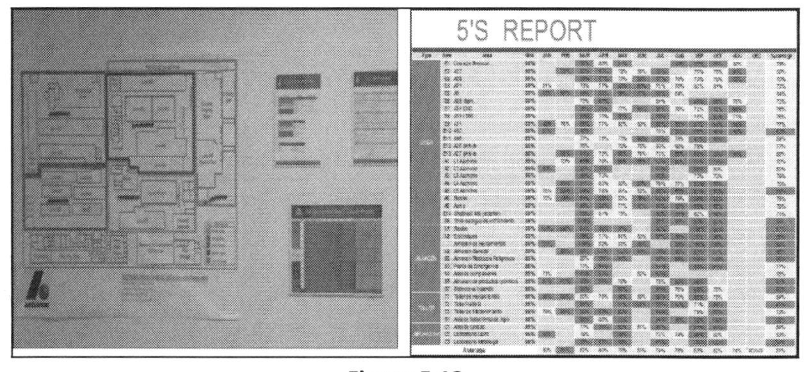

Figure 5.13

Corporate or Optional: 5S boards are a required VOS in Accuride. However, the design is optional in the look and feel of the board. Some are combined with other visually required audit systems such as EHS and Quality. Some basics, though, are that the entire facility and grounds must be incorporated into the overall 5S review.

Figure 5.14

Who Uses It: 5S boards are used by the facilities managers and the auditors conducting assessments. This is where they report their findings.

StdW Balance Sheet

What It Is: Work balance is the review of the standardized process and human interactions needed to achieve both effective (quality) and efficient (speed) results in a work system. Both are optimized in a work balance analysis. Ideally, the balanced system should be designed in such a manner to meet the Takt time of the process. Takt time is the "beat you need to meet" or the rate of customer demand. Accuride has evolved to performing various types of work balance analysis across four differing scopes of work. Generally, we use the format shown in **Figure 5.15**. **Figure 5.16** shows the impact of executing StdW balance across the enterprise. When personnel are impacted by Lean activities, every effort is made to redirect them back into the business leveraging natural attrition and business growth.

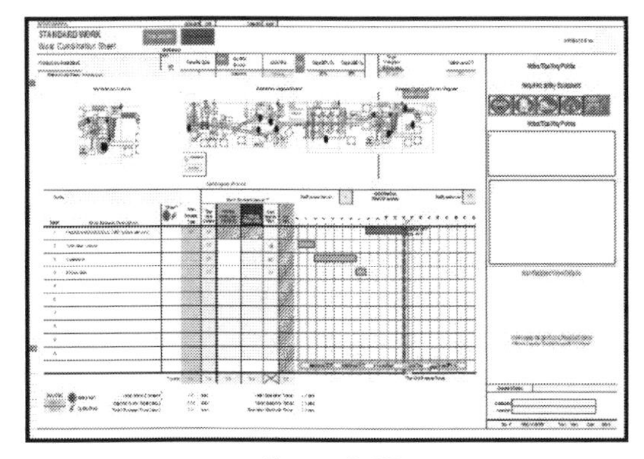

Figure 5.15

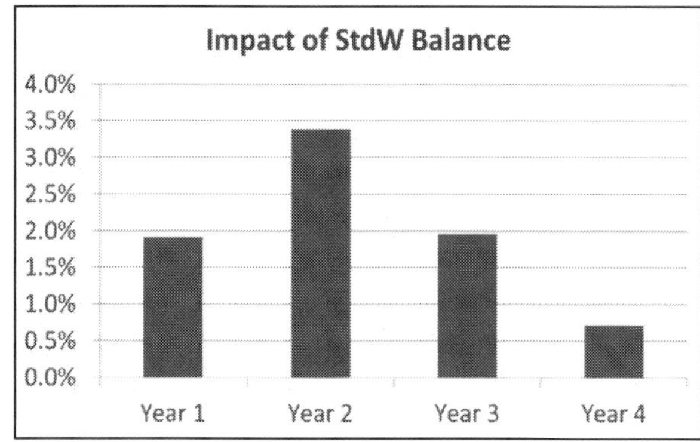

Figure 5.16

1. **Operator Balance:** Evaluation of the value added work, incidental work, waiting time and wasted time when compared to a Takt time.
2. **Machine Balance:** A deeper dive of these four activities when a person is running a machine.
3. **Leader StdW:** Leaders only have so much time in the day in which certain tasks must be performed. Those tasks, and the time it takes to do them, must be identified. While a Takt time may not generally apply, Leader StdW helps to define the core forms of work that are needed to support the teams.
4. **IDL Balance:** A review of StdW and balance for indirect labor positions, such as material handling, maintenance, accounting/financial positions, call centers, etc. This level of review is for roles that have regular routines that are performed continuously. While this type of balanced work analysis is more rare, it has the largest impact to the bottom line. Efficiencies gained here directly impact margin as these tasks rarely involve true valued added activities, which the end customer is willing to pay for.

In itself, StdW is an identified set of work procedures that establishes the most effective and efficient methods and sequences for that process and individual. It is a tool used to best utilize people and machines while keeping the rhythm of process tied to the flow of customer demand (Takt). Each step in a lean process should be defined and performed repeatedly in the same manner. Variations in the process can create quality and/or efficiency problems requiring costly rework or scrap.

Corporate or Optional: Work balance analysis is conducted at all levels across the organization. Business impact is reviewed quarterly. Direct labor reviews for Operator Balance and Machine Balance are conducted iteratively. Most positions have been evaluated at least three times across the company in five years. As Accuride works to grow, the intent is to be able to positively redirect personnel into positions, which would have required someone to be hired into, or to backfill natural attrition. The key metrics impacted include Controllable CPU, Productivity and/or Throughput.

Who Uses It: Work balance is used to help our team members be successful in their positions without overburdening them and/or by having unevenness in their work flow. The analysis helps to optimize the number of personnel needed in the various work functions. It is a critical part of the planning process to ensure effective product costing and business agility.

Transactional Value Stream Maps (T-VSMs)

(See details and insets in Chapter 4)

What It Is: T-VSMs are created *to depict the flow of an information/data creation and delivery process*. Information flows need to be effective (right the first time) and efficient (in a timely manner and when needed). T-VSMs can be drafted in a variety of formats that match the complexity of the process as shown (**Figure 5.17**). Most common are linear maps and/or Swimlane maps. In both, a mini-SIPOC (Supplier, Input, Process, Output, Customer) is typically completed at each step along with key performance data. Common types of performance data could include Lead Time, Queue Time, Cycle Time, %Accuracy or Yield, %Rejects or Return requests for more information.

Corporate or Optional: T-VSMs are required to be conducted on core transactional value streams across the business. The actual format of the T-VSM is adapted to best support the complexity of the process being evaluated. T-VSMs have been conducted across Finance, Supply Chain, Quality Management Systems, Warranty Processing, Human Resources, Inter-Company Sales, Engineering and others. Each functional department exists to provide core functions. Those functions are expected to be as effective and efficient as feasible. Chapter 4 has multiple examples of transactional VSMs and their impact. Remember to note the power of Replication for T-VSMs where one site can construct an initial T-VSM and then cascade it to its sister sites for alignment. This form of “catchball” process for T-VSMs can rapidly align an organization of comment T-VSMs across the business.

Who Uses It: Typically, all indirect personnel across Accuride use T-VSMs to evaluate their key process flows. Internal customers and suppliers are well defined as Cross-Functional Teams (XFTs), which reviews the inputs and outputs across these flows. T-VSMs are also heavily used with external partners, such as customers and suppliers.

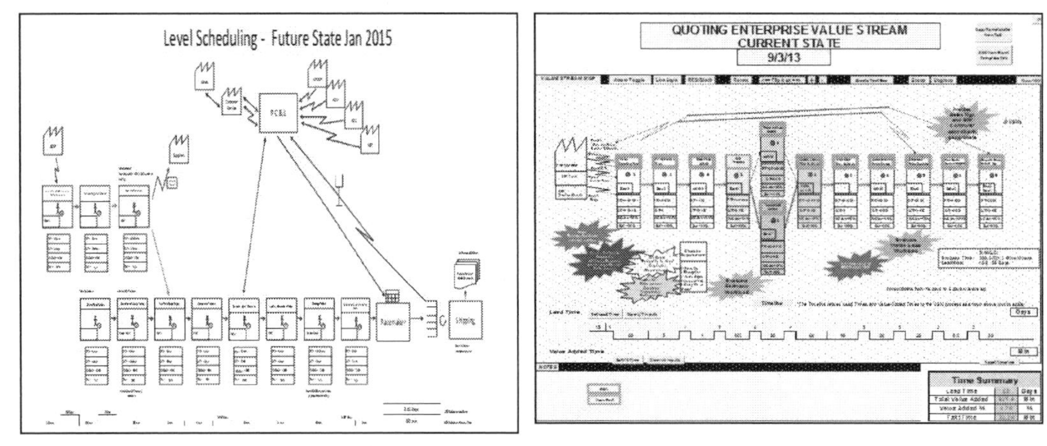

Figure 5.17

Receiving Board

What It Is: Receiving boards are a signaling device that is coordinated between Accuride and its suppliers. A receiving board (**Figure 5.18**) may also be used within the four walls of a site.

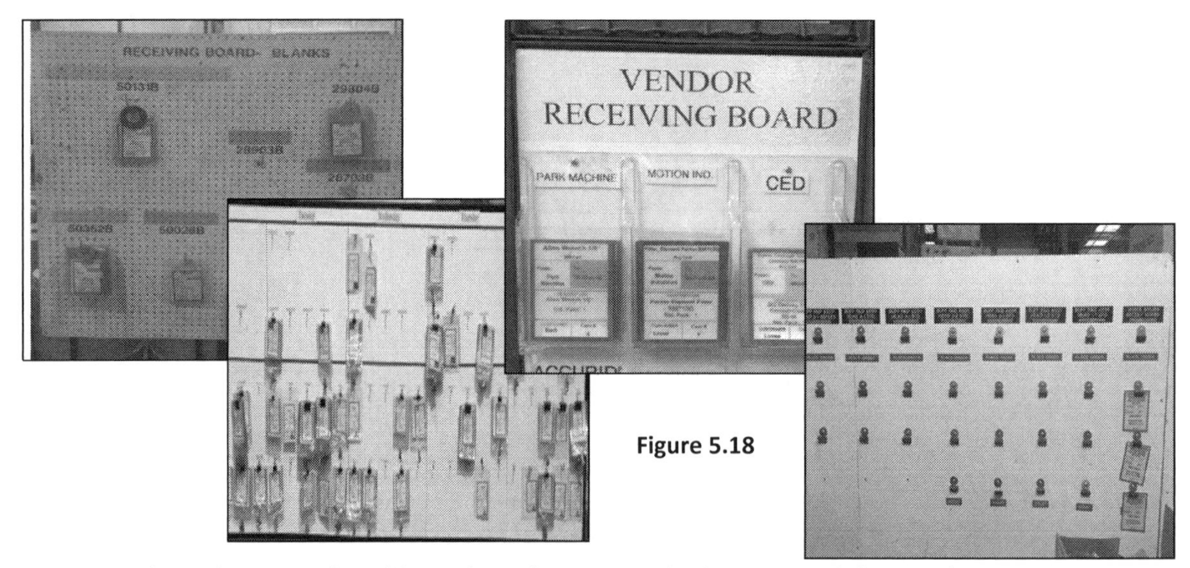

Figure 5.18

- Receiving boards are updated based on the PFEP calculations and demand criteria established between the two parties; whether external or internal.
- Via kanban cards, Fax-Bans, dunnage return triggers or any other pull signals, purchasing confirms the imminent arrival of incoming materials/goods.

- Receiving plans for the receipt of goods and then schedules the timing of the delivery logistics usually by week, day or hour. This allows the receiving team to plan when to have its personnel on-hand to unload and virtually eliminates the cycles of feast or famine at the back door. It largely reduces the mess created by massive unloads that have to be sorted through later and double or triple handling when product has to be dispositioned later. This also assists the delivery teams who are not stacked up waiting to unload.
- Receiving then advises carriers of the delivery timing requirements and secures necessary acknowledgements.
- Kanban cards (now made available when finished goods are shipped) are loaded for receipt into the timing slots and matched with the materials upon arrival. The receiving team now knows how much of what is coming and when. Materials are then delivered with the kanban cards either to Point Of Use (POU) or to an interim location.

Corporate or Optional: At least 80 percent of material spend is managed via kanban cards for receiving. In addition to raw materials, this includes MRO supplies, dunnage materials and any other general products that are included in the overall PFEP. Receiving boards are required and their format can be highly variable depending on the types of products and volumes received.

Who Uses It: Production Control in close coordination with the Shipping/Receiving team uses Receiving Boards. The boards are purely driven by kanban card pulls after downstream products are consumed. A core benefit of the boards is the assurance of on-time delivery of replacement goods while enabling a staggered delivery of goods to balance the flow of the receipts. Besides the reduction of mess, double/triple handling, receipt errors, etc., fork lift needs have been reduced in some areas as a result of this process.

Shipping Board

What It Is: Shipping boards (**Figure 5.19**) are a similar concept to the receiving boards. As pulls are notified by customers via their signaling methods (kanban cards, EDI orders, blanket orders, etc.), a controlled logistics plan is developed to pull, package, stage and ship those items. Shipping lanes are established at larger sites. The carriers are notified for not only the day of pick-up but a time slot within the day. The time slot can be the half day, a specified hour or 15 minute intervals. When the products are loaded, the kanban cards are then returned to the appropriate Changeover Board for replenishment, and the cycle begins anew.

Corporate or Optional: Shipping Boards are required at every site. At least 80 percent of the sold product volume is managed on shipping boards. Their format can be highly variable depending on the types of products, volume shipped and types of carriers used (boat, rail, truck, etc.).

Who Uses It: Production Control in close coordination with the Shipping/Receiving team uses Shipping Boards. The boards are purely driven by kanban card pulls as upstream products are completed. A core benefit of the boards is the insurance of on time delivery of ordered goods while enabling a staggered delivery of them to balance the flow of the outbound goods. Besides the reduction of mess, double/triple handling, receipt errors, etc., fork lift needs have been reduced by 34 percent across the enterprise (see **Figure 9.5**). **Figure 5.20** depicts the "forklift ballet" concept used at one location to optimize single direction handling flow.

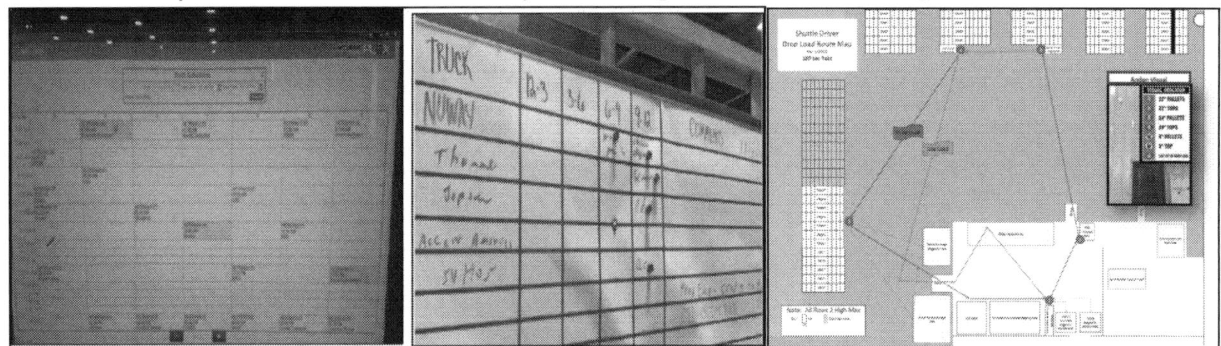

Figure 5.19

Figure 5.20

Tuggers

What It Is: Tuggers are an internal delivery mechanism. An operator's ability to manage their process uninterrupted is key to both process quality and efficiency. By providing them what they need when they need it, they can sustain their focus on the process at hand. Kind of like a pit crew mentality, the operator is the driver of the asset. The rest of the team needs to anticipate the operators' needs and fulfill those needs proactively to minimize downtime. In addition, tuggers are used to take away products, again, based on fill signals, so that the operator does not have to be interrupted to manage with wastes, finished goods, etc. Tuggers work with kanban cards and/or visual signals. Based on the PFEP and/or the rate of consumption of the component, the product is replaced before it runs out so that the process is not interrupted and negatively affected by having to stop.

Corporate or Optional: The tugger concept is leveraged at all production facilities. Tugger design (**Figure 5.21**) is based on the types of products needing to be delivered as well as their frequency. Routes are specifically designed and usually timed as shown in **Figure 5.22**.

Figure 5.21

Who Uses It: Tuggers are meant to support the operator so that their efficacy can be optimized. Based on component usage (via PFEP) the production control team works with operations to determine which items are to be delivered when and in what quantities. Based on the PFEP, a market size is established for how many of each type of product needs to be maintained across the process. This includes raw materials, WIP and FG. A minimum (min) amount and maximum (max) are determined. When a min is encountered, pull signals (kanban cards) are triggered. If a max is encountered, the process is full and upstream activities are discontinued. Teams are redeployed to perform other work.

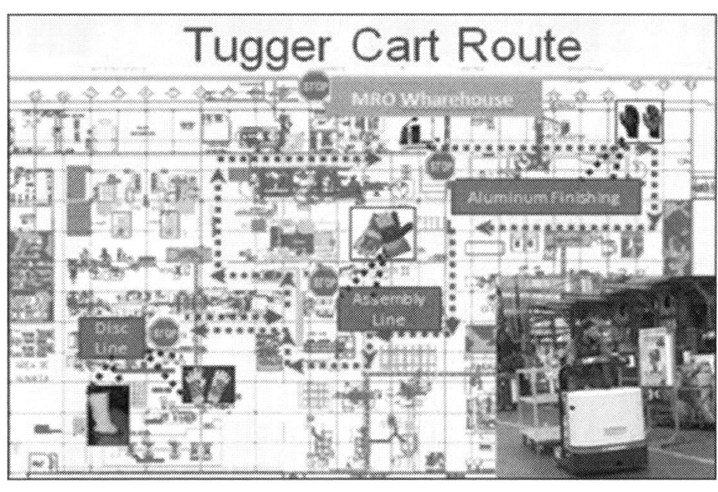

Figure 5.22

Min/Max Signals

What It Is: Based on the PFEP, a market size is established for each type of product specifying how many units need to be maintained in inventory. These levels are determined across all processes. This includes raw materials, WIP and FG. A min amount and max are determined. When a min is encountered, pull signals (kanban cards) are triggered. If a max is encountered, the process is full and upstream activities are discontinued. Teams are redeployed to perform other work. **Figure 5.23** demonstrates the use of min/max levels at Accuride.

AccuLaunch™ is used to determine the footprint areas for these various markets across the processes as well as how to move and handle the product.

Corporate or Optional: Min/max signals are a required VOS in Accuride. The actual signage formats can be different across the sites so long as the basic criteria is easily discernable. This enables the teams to instantly and visually evaluate the health of the process flow and/or if there are immediate concerns that need to be managed.

Who Uses It: Production control works with operations to identify the min/max market sizes based on the PFEP. Engineering helps with footprint layout. An effective use of min/max markets significantly reduces the amount of

Figure 5.23

working capital tied up in the process. Upon establishing effective process flow, the min/max logic in the PFEP can be reviewed for iterative refinement.

Layered Process Audit (LPA) Board

What It Is: An LPA is an auditing process to evaluate the effectiveness of a variety of systems, such as quality, EHS, Lean, etc. All areas of the facility are evaluated through the LPA. A color coding system is used to determine the effectiveness of the system under review. **Figure 5.24** shows one type of LPA board being used at Accuride. Corrective actions are included as a part of the feedback if the results show that counter measures are needed.

Figure 5.24

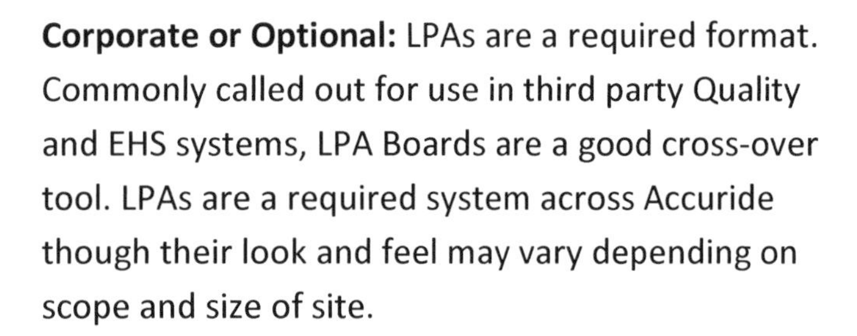

Corporate or Optional: LPAs are a required format. Commonly called out for use in third party Quality and EHS systems, LPA Boards are a good cross-over tool. LPAs are a required system across Accuride though their look and feel may vary depending on scope and size of site.

LPAs have some general required components, such as location of evaluation, color-coded pass/fail results and countermeasures being taken for any gaps identified. Auditors conducting the LPAs are typically not immediately responsible for the area being assessed to ensure objectivity.

Who Uses It: LPA Boards are used by the immediate team that is receiving the feedback as well as by the personnel conducting the evaluation. Site management reviews the findings to ensure that countermeasures are appropriately being taken.

EHS Board

What It Is: The EHS board is a visual awareness summary of safety incidents that occur across the facility. It incorporates both recordable and reported near miss events. Two diagrams show the accumulation of events: The first is on a human figure and the second is within the work area itself. This helps to rapidly provide feedback for areas of training and improvement focus. **Figure 5.25** shows the EHS board at one of the Accuride plants.

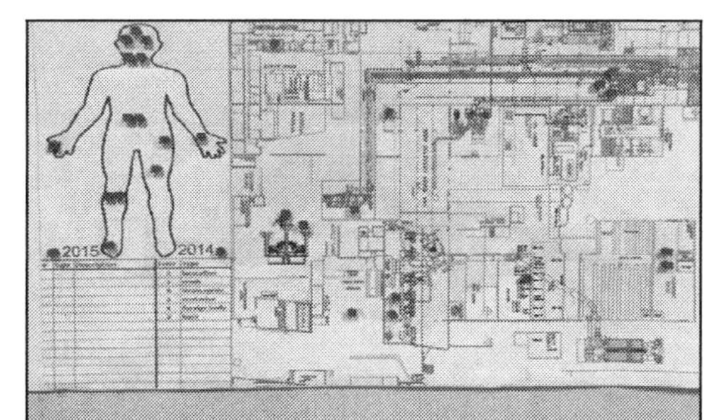

Figure 5.25

Accuride benchmarked this process from Autoliv during a Lean Tour. With an intense focus on the reduction of recordable events and leveraging an EHS Council to share Learning To See and Lessons Learned events, Accuride has significantly reduced its recordable events from the original baseline. One site has had zero recordables for three years and another is on its second year.

Corporate or Optional: The EHS Board format is a StdW VOS across all ACW production locations. Data is rolled up to corporate and reviewed for trends and opportunities to reduce events.

Who Uses It: Facilities Operations, EHS and Maintenance personnel use the EHS boards. While operations reports the events, EHS and Maintenance heavily review the data to formulate methods to design out causals and/or to create more effective systems.

Key Takeaways

1. VOS is an efficient method of information sharing with the team, leadership and the visitors.
2. All key pieces of information are self-explanatory, and the need to ask (and resulted disruption) is minimized.
3. Accuride VOS are classified into two categories – corporate wide standard (required to be followed in a certain format) and optional (tailored to site needs)
4. Each VOS has a supporting procedure with examples.
5. Multiple forms of Corporate VOS target a variety of functional areas.

Chapter Six: Expanding Horizons

So far we have primarily focused on the Lean implementation within the four walls of a plant. Within these four walls we identify a number of opportunities via kaizen bursts in our value stream maps, both production and transactional, to move from the current state to the future state. Even though this helps identify and manage the obvious gaps, an organization is much more than a plant or a singular process. Accuride needed to expand the process to look at the total system. As we looked at the whole system we expected to identify additional constraints to our *mutual* processes. This chapter will look deeper into this side of the systems approach to value stream mapping. Imagine a scenario where an organization has multiple businesses and multiple plants within each of these businesses. The following questions come to mind:

- How do we know that we have an optimum Lean system in place with full visibility to the processes of our internal customers and suppliers to ensure we are able to get the best value out of our inbound supply chain and then deliver that best value throughout the outbound supply chain?
- How do we know that New Product Development (NPD) is effective and efficient?
- How do we know that we have the necessary advance information about our markets so that we are able to respond to changing market conditions in an optimal timeframe?
- How do we ensure that billing processes are aligned to receive payment from the customers on time?
- How do we know that the Supplier or Outside Service Provider (OSP) development processes are efficient so that we are able to both receive and deliver products and services on time?

We need all of the above to happen simultaneously. This symphony needs to be orchestrated across the entire business. The broadening of this scope to include all the sites, sales and marketing, engineering and supply chain elements is called an E-VSM (We've seen other businesses do this with different naming conventions).

The basic concept of the E-VSM is the same as that of any Lean VSM: Manage the flow to reduce the overall LT. Accuride took a "bolt-on" approach to this process; addressing one key system at a time. This enabled us to focus on an outward optimization for these transactional systems. Each progressive iteration identified opportunities for improvement via kaizen bursts. Action items with owners, timelines and process impact were identified. Accuride's F-VSM method was to look three months out; each future state E-VSM had that timing in mind. This made for a very aggressive march through the transactional support systems.

We evaluated the first overall scope of customer order entry to customer delivery. At the time Accuride had three business units with various value streams within them:

1. Wheels
    a. Steel Wheels (North America): Three manufacturing sites, OSPs and a warehouse.
    b. Aluminum Wheels (North America): Three manufacturing sites, OSPs and a warehouse.
    c. Steel Wheels (Europe): One manufacturing site. Self-contained.
2. Wheel End Products
    a. Drums: One manufacturing site and a warehouse.
    b. Hubs & Hub Assemblies: One manufacturing site, OSPs and a warehouse.
    c. Slack Adjusters: One manufacturing site and a warehouse.
3. Brillion Iron Works
    a. Peters Processing: One manufacturing site. Self-contained.
    b. Gabler Processing: One manufacturing site. Self-contained.

Using the E-VSM for Steel Wheels (North America) as a *model* below in **Figure 6.1**, the sites are positioned in the middle. The far left shows the incoming supply chain and any component transformation while the far right shows the OSPs and delivery models.

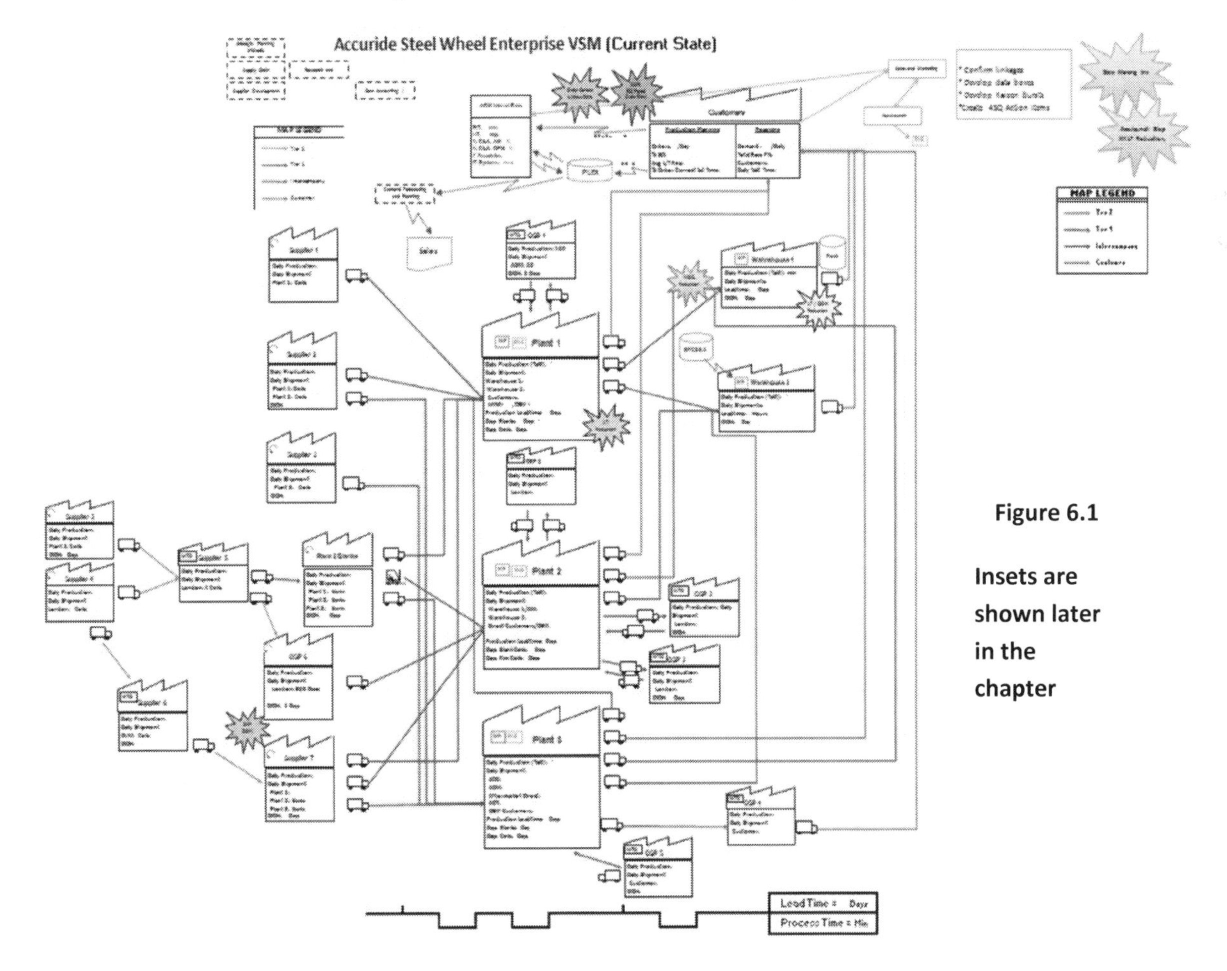

Figure 6.1

Insets are shown later in the chapter

At a glance it can look overwhelming. However, when stepped through logically, it is powerfully revealing. It follows the same logic of a regular VSM with the customer at the top right corner, suppliers at the left, operations in the middle and OSPs (if there are any) at the right. The key difference is that all of the plants are now treated as one entity. The impact of the slowest or highest cost producing site becomes glaringly obvious. We found it critical that each of the site's internal VSMs be as optimized as possible prior to completion of the enterprise level review. For simplicity, we have partitioned the E-VSM into five key sections:

1. Customer order placement within the organization and its communication to the plants.
2. Inbound supply chain (suppliers) delivering raw materials to the plants.
3. Plants producing the products.
4. Plants providing semi-finished products to be processed by OSPs.
5. Plants delivering to the customers directly or via warehouse/final processing center.

Customer Order Processing and Communication to Plant

Customer orders are received from the production planning function through various channels, such as Electronic Data Interface (EDI), email or fax and are consolidated via an internal sales function. The output of this process is fed into all the plants. An internal sales function consolidates the orders (**Figure 6.2**). The data box for internal sales function contains the following information:

1. P/T: Min (Process Time)
2. L/T: Days (Lead Time)
3. %C&A: Percent (By key Customer Type (Complete & Accurate)
4. Quantity Associates
5. IT Systems: PLEX

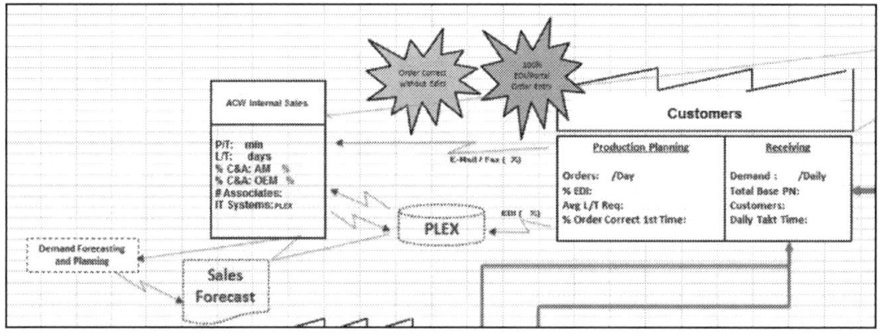

Figure 6.2

Some of the terms are:

P/T (Process Time): The time commonly required to complete the process. This may vary, but needs to be representative of the process. If it is too hard to lock in a specific time, use a range. For example, it may take 5-25 minutes to complete a report. If there is a huge variation, due to missing information, this would be reflected in the %C&A (Complete & Accurate; defined below) from the previous process. Process Time should be based on an average of the experienced processing time, the rarer "what-if scenarios" should be excluded.

L/T (Lead Time): Time that the information/material waits before the first action is taken. The unit of measure may be the same or different from the process time, depending on the time range. For example, the TPS report sits in queue prior to being processed for zero to two days. Lead time should be based on an average variation in waiting time, the rare "perfect storm event" should be excluded.

%C&A (% Complete & Accurate): Defines the quality of the input, provided by the previous process steps. In other words, the percent of time that you don't have to ask the upstream process for correction or clarification prior to completing the task at hand. For example, Report X has the correct data table pasted into it 95 percent of the time prior to being received.

%AOP (% Availability Of Personnel): The percent of time that employees are "truly" available to work in the value stream. This number may not be 100 percent if the department completes a variety of tasks. For example, in a 40 hour week there may really only be six hours available to work on process X (15 percent AOP) because 32 hours are devoted to adjusting process Y.

Inbound Supply Chain Delivering Raw Materials to the Plants

Figure 6.3 illustrates the concept of delivery of supply chain components; from raw materials to and through sub-assembled goods. This model shows the supply of raw materials and key components to differing sites. The data in the box for suppliers might detail:

1. Daily Production
2. Daily Shipment in Units shipped
3. DIOH Days

Figure 6.3

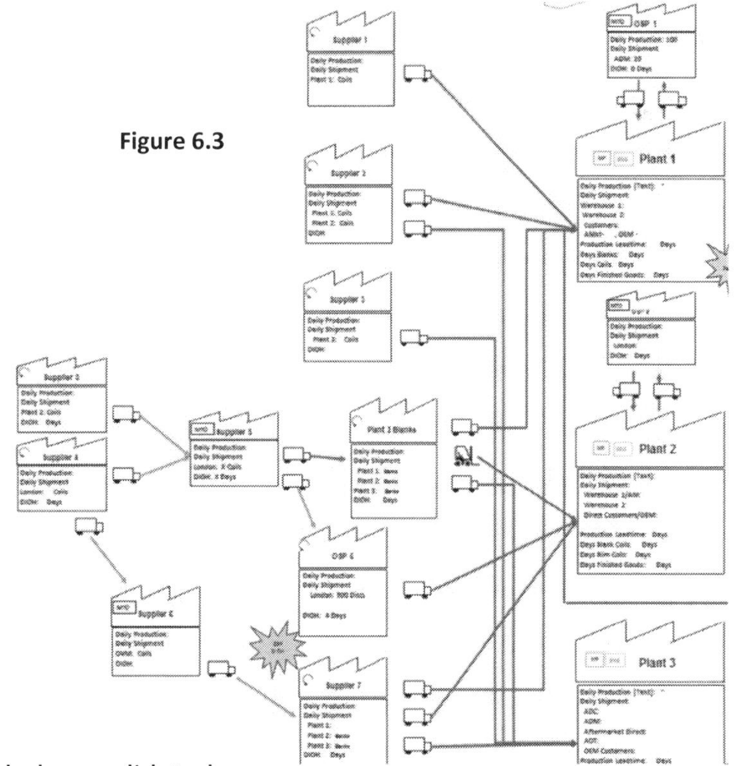

This gives the assessing team an opportunity to look at the supply chain from its overall context. A look at the inventory gives an opportunity for more cost reduction. This value is further enhanced once the suppliers and sites are translated onto a geographical map along with the quantities flowing across them on a daily basis. This exercise gave us the opportunity to look at the inbound supply chain differently, enabling be for system optimization. These included consolidated milk runs with a delivery cadence, drop-off/pick-up synergies, load usage optimization, etc. This was executed across the entire external delivery chain.

Plants Planning and Producing to Meet Customer Requirements

The producing sites are the center of the E-VSM in **Figure 6.4**. Material planning notes may be included at the top of each box for each site. This planning is fed from the inside sales, customer kanbans (as applicable) and the sales forecast (based on product type). Inside the data box, each plant describes the following information:

1. Daily Production (Takt): Seconds
2. Daily Shipment:
   a. Each Warehouse Identified
   b. Customers (By Type)
3. Production Leadtime: Days
4. Days for Key Component Streams: Days
5. Days for Finished Goods: Days

Figure 6.4

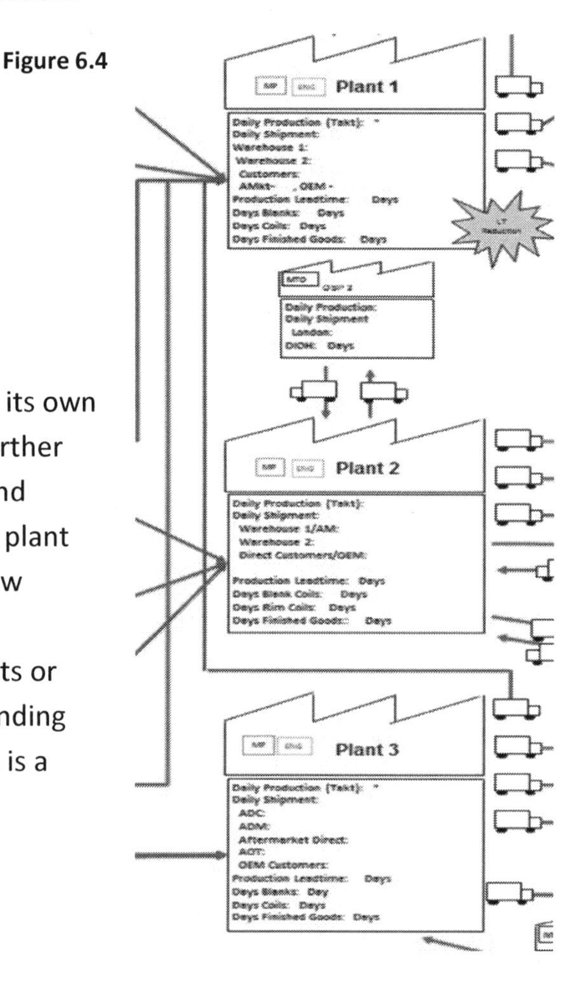

The data box for each of the plants is supported by its own VSM. It gives an opportunity for the site team to further refine its internal flows with respect to lead time and inventory reduction opportunities. At this level the plant boxes also include capacity data. The capacity is now compared to the forecast. A percent utilization is established to understand the (potential) constraints or overages. Capacity percentage is color coded depending upon the level of utilization. The table in **Figure 6.5** is a conceptual example.

	85-90%
	Below 85%
	Above 90%

ESW Production Capacity Utilization Percentage (7 Day)

	Blanker		Spinner		Disc		Rim		Wheel Assembly			
	Capacity	% Utilized	Capacity	% Utilized	Capacity	% Utilized	Capacity	% Utilized	Capacity	% Utilized	2016 Forecast	
Plant 1	–	–	1,600,000	94%	2,000,000	75%	1,800,000	83%	1,600,000	94%	1,500,000	Plant 1
Plant 2	10,000,000	50%	1,500,000	67%	1,600,000	63%	1,100,000	91%	1,100,000	91%	1,000,000	Plant 2
Plant 3	–	–	2,800,000	89%	4,000,000	63%	4,000,000	63%	2,800,000	89%	2,500,000	Plant 3
ESW TOTAL	10,000,000	50%	5,900,000	85%	7,600,000	66%	6,900,000	72%	5,500,000	91%	5,000,000	ESW

Figure 6.5

The capacity example provided shows a seven day running model. This accounts for the maintenance and other activities that have to be completed on a weekly basis. Similar tables can be prepared for six and five day run rates depending on work schedules. Based on growth or contraction, this gives the business a way to manage capital expenditure planning with respect to balancing assets, minimizing bottleneck impacts and further reducing overall product cost.

Plants Getting the Semi-finished Products Processed by OSPs

OSPs are a critical aspect of process flow as they can significantly impact LT due to their own location and internal process efficiencies. They may also be customer directed for sub-assembly processes. Sometimes OSPs can act as a buffer to accommodate significant fluctuations in the customer demand; such as products coming from overseas. As shown in **Figure 6.6**, an OSP data box might include the following information:

1. Daily Production
2. Daily Shipment Units shipped
3. DIOH Days

Figure 6.6

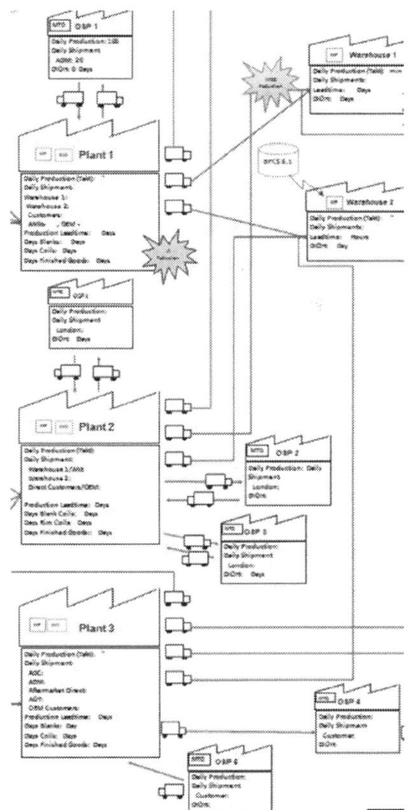

The value of including OSPs in the E-VSM is similar to that of including suppliers in the VSM; the optimization of the relationship that delivers economic value to the customers. Another important benefit is the inclusion of DIOH since this inventory is typically owned by Accuride and needs to be processed optimally.

Plants Delivering to the Customers Directly or via Warehouse/Final Processing Center

Depending upon the nature of the product, they are generally either directly shipped to a customer or to a warehouse for short term storage and subsequent shipment. At Accuride, one of the warehouses acts as a vertically integrated OSP and provides further value add for a dedicated customer. This is common in the OEM world where product is requested in a specified sequence. Ultimately all the products end up at the customer's locations via their receiving teams. In a VSM or E-VSM, this is at the other end of the customer's data box and completes the loop, which started at the left side of the production planning box. For warehouses, DIOH is the most important metric directly linked to the plant lead times via the PFEP (Discussed later in this section). **Figure 6.7** models this part of the E-VSM.

Tying it all Together; What we Learned

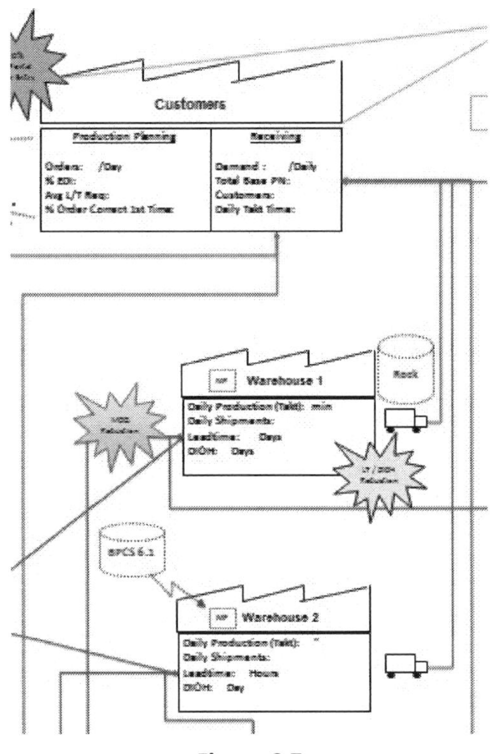

Figure 6.7

As mentioned earlier, the scope of this E-VSM is to cover all the processes involved from the customer order entry to the customer delivery. The E-VSM cross-functional team need to have the necessary decision making team members involved in the process for it to be effective. The team needs to have permission to try things to see if they will work; this approach is conservative and has not adversely affected customers even though we've had an occasional upward inventory tick. Critical team members include those who have ownership and decision making authority for:

1. Plants
   a. Production Control/Supply Chain
   b. Operations
   c. Lean
2. Warehouse: Director/Management
3. Corporate
   a. Master Black Belt/Industrial Engineer/Facilitator
   b. Supply Chain
   c. QLMS
   d. Engineering
   e. Sales/Marketing
   f. Inside Sales

A large discovery was the additional identification of opportunities to further reduce and optimize working capital across the enterprise. Efficiencies gained in the transactional teams enabled the delivery of more timely and accurate data. Logistics planning itself went through a full overhaul via the leveraging of third parties that helped to optimize the delivery patterns and loads. There was an overarching clarification of process capacity and capability. Excess inventory buffers were highly magnified and substantially reduced. As the plants continued to drive internal cycles of improvement (especially with changeover time efficiencies), the impacts were seen within the enterprise results. This enabled the overall business to further reduce lead times and Minimum Order Quantities (MOQ) both to ourselves and to our external customers and suppliers.

Even though the sites had ***independently reduced their lead times by an average of ~50 percent***, the E-VSM evaluation added onto the entire organization ***another layer of lead time reductions*** ranging from ~40 percent in the Warehouse to 3-25 percent across the various product lines (**Figure 6.8**). The warehouse saw the largest impact due to the transparency at the sites. This was achieved by just managing the ***transactional flows*** at the enterprise level. Remember, very little to no value add is being performed across these processes.

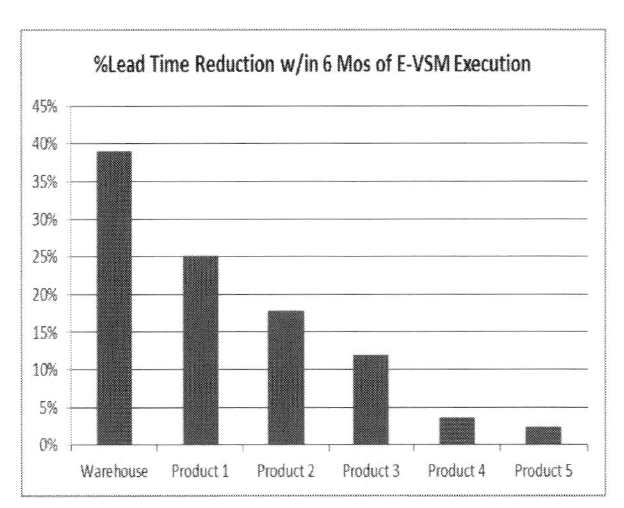

Figure 6.8

Impact of E-VSM to the Business/LT, Reduction/Logistics Impact, etc.

Since time is money, what was the impact of taking this ***up*** to the next iterative level via the E-VSM? *Within six months of execution*, Accuride experienced a net reduction of ~3.9 percent of its total COGS.

Keep in mind that this is now way beyond "Lean Manufacturing." The site-based key value streams have been optimized to some degree. And while there are still cycles of improvement, these are more along the lines of refinement. The E-VSM effort is with team members that are culturally experienced and have the AccuPride™ attitude that there is more room for improvement.

Next Steps...

As mentioned earlier, the current E-VSM scope focused on optimizing key operations from order entry to the customer delivery for the business. However, there are additional transactional processes that can be further enhanced to optimize business performance, such as NPD, Customer Relationship Management (CRM), supplier development, working capital management, etc., all with the eye towards improving Accuride's cash cycle. These are a few of the next areas of focus after we have achieved sustainment of the current flows:

Growing the Depth and Breadth of the E-VSM

Products are classified as Runners, Repeaters and Strangers depending upon their volume of consumption. The concept is that not all products need to be treated equally. Runners, the highest volume products, consume the most resources. These can be optimally controlled via kanban. Strangers, rare cats and dogs, are typically made to order products, and Repeaters have some frequency but are in between. The teams are now taking a "3D" approach to the E-VSMs and further tailoring the flows at these levels.

The teams have looked at the overall product portfolio and have created a representative E-VSM. Evaluating the E-VSM separately for runners, repeaters and strangers gives a totally different perspective. Our first pass analysis has yielded some very interesting results: Runners had higher than expected inventory for finished products, repeaters had high WIP and interestingly, strangers seemed to be quite in line with expectations.

As the sites review the findings and re-evaluate their PFEPs and TTRs, they are introspectively reviewing their markets for additional opportunities. One approach is to gradually reduce the inventory for the runners while looking for reductions in repeaters. These actions would further reduce lead time and working capital. Once stabilized and evaluated for effectiveness, the business should be able to outwardly adjust its lead time guidance to the customers.

Future E-VSM Scope

Finance: Even though each transactional team has the responsibility to introspectively improve its own processes and deliverables, from an E-VSM perspective, finance integration is critical with respect to reducing the order to cash cycle.

New Product Development (NPD): Inorganic growth via successful NPD is critical to business growth. The pun of "reinventing the wheel" is real for Accuride. NPD can become more effective (what the customer wants) and efficient (time to develop). As new items are developed, they need to be introduced into the sites for product samples and testing of manufacturing processes without being disruptive to the normal flow of operations. Engineering sample requests are a part of the E-VSM and within the AccuLaunch™ process. Treated as Strangers (Make To Order), these can have an impact on a site's capacity due to special processing needs.

By providing visibility to the number of new products planned at the plant level and their due dates, the pipeline is visible to all of the stakeholders in the enterprise. With periodic review, the status becomes clear and the necessary priority can be assigned to ensure that the business goals are being met by proactively identifying and removing the constraints.

Sales: By integrating the sales function, the E-VSM team is able to provide visibility to any slow moving inventory. The sales team can then determine the next best steps that would be needed. Special promotions, bridge activities and other cross-functional actions can be managed through these flows.

Supplier/OSP Development: In order to remain competitive, businesses focus on the supply chain because it accounts for a significant portion of COGS (>60-70 percent). We continue to partner with suppliers on effective and efficient means of communication and provide some of them with Lean support so that they, too, can optimize their flows to us. By integrating Supplier Development there is a visibility to assess the impact within the overall map.

Regardless of the inclusion of the necessary transactional teams, at some point, the E-VSM will self-optimize. After that, ***the next frontier lays with sustaining the new processes and ensuring that they avoid the inefficiencies that had been created in the past***. How will the operations and transactional teams develop future production and business processes to operate as effectively and efficiently as possible to prevent these inherent losses from being incurred in the first place?

Key Takeaways

1. E-VSM goes beyond the four walls of the facility and into the overall enterprise to include inbound and outbound supply chain and transactional components.
2. The initial scope of the E-VSM at Accuride covered the customer order entry to the customer delivery as the product flows across the enterprise.
3. An integrated view puts the plant capacity planning to meet the forecasted demand in perspective.
4. Transparency of the process results in significant value recovery in terms of reduced inventory, lead time and logistics expense.
5. Additional opportunity to expand the E-VSM scope may be uncovered with a "3D" analysis in refining lead times by product classification (Runners, Repeaters and Strangers) and then assigning accurate lead times to further reduce the inventory via accurate PFEP.
6. By integrating other transactional processes, such as new product development, sales, finance and supplier development, further EVSM benefits can be realized.
7. The next frontier lies in sustaining the new processes including the developing of effective and efficient processes so that more optimal cost benefits can be attained sooner.

Chapter Seven: Harmonizing the Lean Systems

As discussed in Jim Womack and Daniel T. Jones' book *Lean Thinking* (**Figure 7.1**), there are five principles to Lean thinking:

1. Precisely **specify value** for specific product.
2. Identify **value stream** for each product.
3. Make value **flow** without interruption.
4. Let the customer **pull** value from the producer.
5. Pursue **perfection.**

Figure 7.1

Let us look at each of these principles and how they were put into practice at Accuride:

1. **Specify Value**

   Accuride actively identifies value in terms of the voice of the customer. All Accuride sites are certified to ISO/TS 16949, (IATF 16949:2016). Accuride not only makes the extra effort to understand customer expectations in terms of products and services (delivery and post-delivery), but also actively works with customers to establish pull systems (wherever they will let us!). Our sales and engineering teams proactively work with our customers to refine the meaning of value. While the ultimate customer defines the value, the customer may not know the cost implications of some their own expectations. A true partnership with customers will help uncover these cost opportunities, which may not be readily visible to by the customer.

2. **Identify Value Stream**

   Chapter 4 is devoted to the concept of value streams. The main concept is to keep refining the process from a value add perspective to reduce the lead time (which is viewed as waste). This focus has moved from production processes to transactional processes and now to the enterprise wide VSMs with clear visibility to the next steps.

3. **Make Value Flow**

   Flow is Lean. With discussing flow, it is important to distinguish between three terms: constraint, bottleneck and pacemaker. A system has multiple processes where each process goes through multiple steps. There is a demand on each of these steps, and each step has a certain capacity.

A bottleneck is a process step that has capacity that is less than or equal to the demand placed on it. A constraint is a limiting factor, or the weakest link, in meeting the demand. Shown in **Figure 7.2** is a high level summary of Goldratt's *Theory of Constraints*.

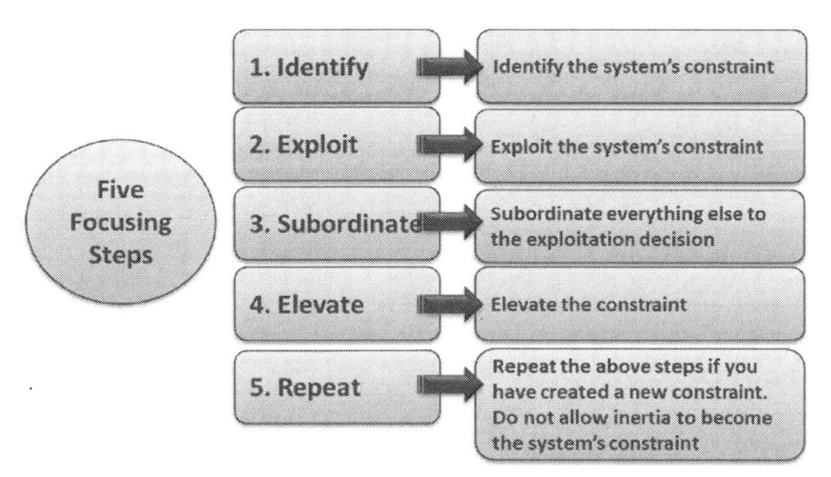

Figure 7.2

Now that we have distinguished between a constraint and a bottleneck, where does the pacemaker fit? *The pacemaker is the step after which the product or service flows to the customer.* The schedule is typically provided only to the pacemaker and thereafter the product and service flow to the customer. All upstream processes respond to the demand of the pacemaker. The pacemaker may or may not be a constraint. At Accuride, OEE is measured only at the pacemaker.

If the process is not able to meet the Takt time, the underlying impediments to that flow need to be understood and removed. Once Takt is achieved, further improvements can be made to reduce the per unit cost. Part of this analysis requires a verification that the products being produced are actually what are needed by the customers.

1. **Let the Customer Pull**

As mentioned in the section for value, customers only buy what they need. Accuride actively works with its customers and plants to put as many items on a pull system as possible. This helps both entities. The system transparency removes planning, guesswork and uncertainty. Need for safety stock can be dramatically reduced. With the E-VSM there is further refinement in the the lead time for planning, resulting again in a further reduction of inventories. Many organizations are good at just-in-time supply but not at just-in-time production. It is this gap that can be be minimized by working closely with the customers.

2. **Pursue Perfection**

There is an old saying that no one is perfect. However, one can always have a relentless pursuit of perfection. At Accuride, there are baselines established for the key processes and then continual improvement is pursued with the intent of making them more effective (getting closer to the intended outcomes) and more efficient (using minimal resources in terms of efforts, time, space and errors/scrap). In order to pursue perfection, one needs to look at all the critical inputs to any process:

Materials	Information	Machine	Tools
Method	Approach	Environment	Measurement System
People/Skills			

All of these need to be evaluated in terms of:

a. Adequacy: Is what is being provided enough, and how do we know that it is?
b. Effectiveness: Are we getting the intended results? If not, re-evaluate adequacy.
c. Efficiency: Are we able to deliver the intended results with minimal but adequate resources? This involves refinement of adequacy.

Accuride's Lean system has been implemented in a manner that continually evaluates adequacy, effectiveness and efficiency. The teams are regularly trained. Frequent communication helps share the new information/knowledge across the cross-functional teams. Let us now tie together what has worked at Accuride so well.

Accuride's Lean System Overview in an AME Format

The sections below are summaries from applications across four of Accuride's AME Award winning sites: Henderson, KY; Erie, PA; Rockford, IL; and Monterrey, Mexico. Many of the figures are referenced in other sections.

Accuride's QLMS Council

Accuride's QLMS Council consists of all the plant Lean leaders, warehouse leaders, functional team leaders such as IT, HR, Engineering, Supply Chain and Sales. A sample meeting agenda is shown in the **Figure 7.3**.

AGENDA

Time	Topic/Discusssion Item	Who	Time
10:00	Attendance. Roll-call.	VP	0:03
10:03	**QLMS.** Calendar: Review calendar	Round Robin/MBB	0:01
10:04	**Status:** Hoshin Kanri X-Matrix Status Update...Overview...	VP	0:10
10:14	**Status:** Transactional VSMs...round robin; all sites	Lean Mgrs	0:00
10:14	**F2F Planning/Qrtly Lean:** Logistics for Lean F2F	VP	0:15
10:29	**MBB/IE:** Current 4SQ Status	MBB	0:10
10:39	**KPIs.** Review of key data results. Review trends	VP	0:05
10:44	**Next Steps.** AccuLaunch - general status	MBB	0:05
10:49	**Lessons Learned. Plant B**	Lean Mgr	0:05
10:54	**Accuride News:** Sales Leveraging of Lean Systems w/Customers	VP	0:05
10:59	**Next meeting:** Date, Location	VP	0:01

Figure 7.3

Hrs: 1:00

The QLMS meets monthly for an hour and has a standardized agenda, which is timed literally to the minute. Attendance is taken as well as an on-time start and end. Regular features include:

1. Performance status to Hoshin Kanri X-Matrix. Reviewed in a "round robin" manner where each plant provides a one to three minute status update to the topic in review.
2. Functional team reports from: Master Black Belt, IT, Supply Chain, Engineering, Quality, etc.
3. Calendar review of upcoming events: third party audits, customer visits, etc.
4. Sharing of enterprise-wide KPI performance: Gap analysis and next steps.
5. Sharing the Lessons Learned by a pre-selected site or function. A rotational calendar is developed annually.
6. Any key topics.

Quarterly Lean Review

The Executive Council reviews the status of Lean activities across the corporation once per quarter. These reviews may be at a site location or via WebEx. The council uses these on-site visits to learn from each other and further affect standard work. The Executive Council, key functional leads, Lean management and various team members are present at the review. Each site reports to their Lean status on key value streams and KPIs. This review re-affirms the expectations and assigns priority and responsibility for any identified gaps. **Figure 7.4** shows a typical agenda for a quarterly Lean review.

EST	CST	Topic/Discusssion Item	Presenter	Time (Hr:Min)	Web-Session
10:00	9:00	**Kick-Off w/CEO**	CEO	0:05	**Web Session Starts**
10:05	9:05	Lean Council Overview	VP - Quality, Lean & EHS	0:10	Ongoing
10:15	9:15	Transactional: Finance	CFO	0:07	Ongoing
10:22	9:22	Transactional: Plex/IT	CTO	0:15	Ongoing
10:37	9:37	**Steel: E-VSM.**	Team Leader	0:10	Ongoing
10:47	9:47	Steel: Plant 1	Director of Operations	0:05	Ongoing
10:52	9:52	Steel: Plant 2	Director of Operations	0:13	Ongoing
11:05	10:05	Steel: Plant 3	Director of Operations	0:05	Ongoing
11:10	10:10	Steel: Plant 4	Director of Operations	0:10	Ongoing
11:20	10:20	**Aluminum: E-VSM**	Team Leader	0:10	Ongoing
11:30	10:30	Aluminum: Plant 1	Director of Operations	0:10	Ongoing
11:40	10:40	Aluminum: Plant 2	Director of Operations	0:10	Ongoing
11:50	10:50	Aluminum: Plant 3	Director of Operations	0:05	Ongoing
11:55	10:55	Warehouse: E-VSM Overall	Director	0:10	Ongoing
12:05	11:05	**Foundry: Rockford. E-VSMs wrt Drums, ASA, Hubs**	Director of Operations	0:15	Ongoing
12:20	11:20	Foundry: Rockford. Site Details	Rkfd Team	0:05	Ongoing
12:25	11:25	Closing Remarks/Wrap-Up	CEO	0:05	Ongoing
12:30	11:30	finish Lean review			

Figure 7.4

Accuride's Lean System - Details

Section I: Policy Deployment Process

Section 1.1: Management System

Policy Deployment

The Accuride Executive Committee (EC) starts with a Strategic Vision. With this vision in mind, the EC regularly reviews and issues strategic priorities based on Voice Of the Customer (VOC), business objectives and environmental conditions. These priorities are then communicated across the organization in a variety of venues including plant manager meetings, the corporate newsletter, Town Hall meetings with the CEO and other such activities. In this manner, all Accuride employees are encouraged to learn the objectives of the business.

The EC is responsible for the execution and attainment of the strategic priorities. Results are reviewed quarterly, and the EC meets monthly to review progress against the planned tactics. A standardized metrics system of KPI is reviewed regularly. Key results are managed at every site on a daily basis on Glass Walls: a centrally located status communication center.

Management Approach to Achieve Policy Goals and Strategies

The strategic priorities are embedded within the personnel performance objectives of the Accuride leadership team. Those are then rolled through to their direct reports. A mid-year assessment is conducted Accuride-wide on the performance status of all personnel and their objectives.

Hoshin Kanri (called the X-Matrix) is utilized at the functional level for the execution of the QLMS. The X-Matrix has four key sections: Strategies, Initiatives, Tactics and KPIs. This is detailed in Chapter 2.

As processes are executed, KPIs are reviewed monthly. These KPIs include Safety, Quality, Productivity and Cost. Different KPIs are reviewed at relevant times across the year. They are shared in multiple areas, including the Glass Wall and Cell-based Kiosks as shown in Chapter 5.

Operations and Office Approach for Cascading Goals

All sites use a Glass Wall for the clear communication of the status of its goals. With this type of transparency, every team member has cross-functional responsibilities for the transactional processes required to support the facility. Accuride manages its Production Control, Engineering, Finance, Human Resources, IT, Purchasing and Environmental Health and Safety

(EHS) using Lean methods. Accuride's corporate function supports other transactional areas, such as Customer Service, Sales and Marketing and Legal. This blended leadership team reports on the status and results of its goals and deliverables in a monthly Management Performance Review (MPR) as well as in a quarterly Lean review. All sites leverage an integrated "walk-around" several times a day to proactively assess the process status identified on the VOS boards.

Continual Improvement (CI) Program to Achieve Policy Deployment

Accuride's CI program is driven by the X-Matrix, which is based on Accuride's strategic plan. The plan outlines the priorities for Sales, Operations, Quality, Safety and Technology. From that plan, an aggressive CI program is leveraged to support goal attainment:

1. **Annual** establishment of the Lean and Quality Management System (QMS) X-Matrices based on corporate strategies.
2. **Annual** management **review to ISO/TS 16949** (IATF 16949:2016) systems criteria.
3. **Annual** self-assessment to the Waddell Lean survey.
4. **Annual QLMS skills growth** review based on a suite of approximately 30 QLMS skill sets. Each QLMS Council is assessed to these skills.
5. **Annual VOC review** (corporate level) based on key customer types: Original Equipment Manufacturer (OEM) and Aftermarket.
6. **Quarterly Lean Reviews.** Each site and functional team reports on the Lean status. Team tours provide the facility with feedback. A traveling trophy is awarded to the site that has made the most change in baseline performance from quarter to quarter (**Figure 7.5**).
7. **Monthly QLMS.** Sites take a turn in providing a Lessons Learned to the Council.
8. **Transactional process "replication."** Each site models one to two key transactional processes at their location. They then take the improved process and share it with the other locations. The sister sites can then leverage the process and incorporate relevant improvements. This reduces the lead time of fully constructing T-VSM flows for similar processes and rapidly enables StdW for shared transactional processes.

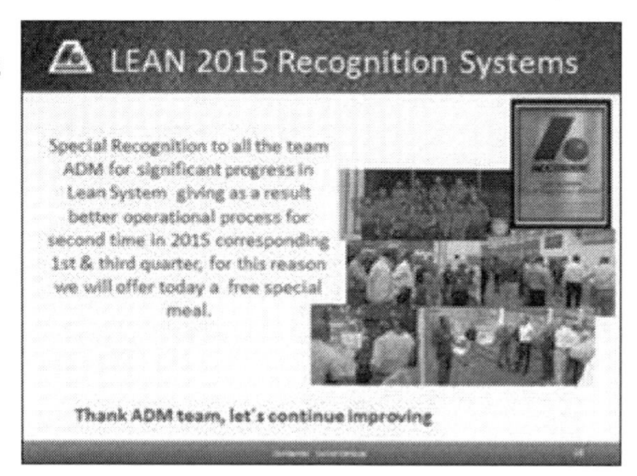

Figure 7.5

Context Page from Monterrey

Role/Relationship of Management and Associates in Achieving Goals and Objectives

Accuride's approach is one of servant leadership. It is the leadership team's responsibility to empower and enable associates to accomplish their goals and objectives. One of the key systems in place to accomplish this is through multiple walk-arounds that are conducted every day. This enables the leadership to be personally available to the teams and to support the process as results unfold. Production leaders literally review the board status every hour. They record actions if targets are not being met. This supports a message of engaged awareness to the team. The leadership team works to remove barriers and find a way to "say yes" when working to resolve an issue. With this approach, the teams have continually improved upon their ability to meet the business' goals and objectives.

Utilization of StdW

Accuride utilizes the Plex Enterprise Requirement Planning (PERP) system to manage its raw materials, Work-In-Process (WIP) and Finished Goods Inventory (FGI). Operations management is then guided via ACW's ISO/TS 16949 QLMS (IATF 16949). Transactional teams also leverage the QLMS system for related policies and procedures. There are more than 20 forms of VOS processes embedded across Accuride. Formal StdW for employees are evaluated at four levels: Operator (Op), Machine (Mach), Indirect Labor (IDL & Overhead) and Leader StdW.

Outlining Expectations and Follow-Up With the Teams

The system of outlining expectations and conducting follow-up occurs across several levels of the business:

1. Leadership team's goals and objectives are aligned with the strategic priorities via the performance goals and objectives setting process.
2. Site initiatives and supporting tactics are developed via the X-Matrix.
3. Performance to the objectives is reviewed monthly and quarterly via MPR and Council meetings.
4. A variety of VOS boards are used to communicate immediate needs and concerns. These boards have a built-in method of communication to show progress on implemented actions.
5. There are three different types of tools used to manage improvement communications depending on the level of scope and complexity:
   i. "Easy:" 5-Why. A one-page format tool used for quick communication and resolution.
   ii. "Medium:" 4-Square (4SQ). A project management tool with defined action items, owners and timing.
   iii. "In-Depth:" 8D Corrective Action/Preventive Action (CA/PA). A detailed problem-solving process used internally and externally to resolve identified issues.

Section 1.2: Human and Organizational Development

Approach to Training and Organizational Development

The Training Program is based on the competencies that each associate requires to be successful in their role. The Competence Model has two phases:

1. Development of the competencies needed for each position.
2. Evaluation and certification of the associates against those competencies.

The main process steps include:

1. Training and assessing level of knowledge.
2. Qualifying the skills for the operation.
3. Evaluation of performance.
4. Achievement of certification.

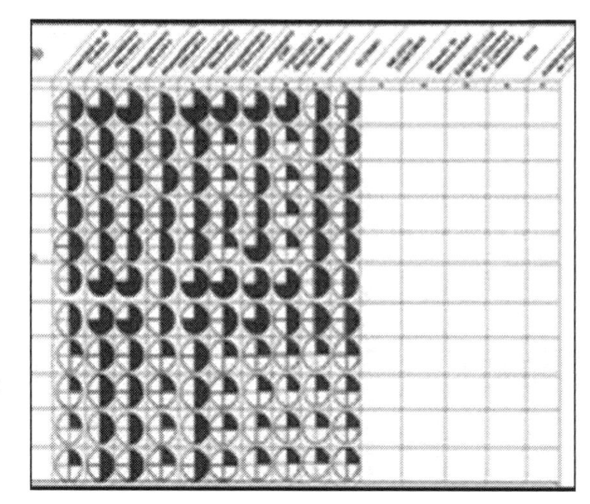

Figure 7.6

Some sites follow a Skills Matrix as shown in **Figure 7.6**, which has four levels of progression. This sample does not show the employees' names or photographs, which is a common format. Developing leadership skills is a basic tenet of the Continuous Improvement process. As a part of the Team Leader certification process, potential Team Leaders are often evaluated with a 360° performance review before they are accepted into a Leadership training program. Accuride helps its associates to develop their Career Plan because we believe in our talent and in developing their skills.

Associate Engagement and Morale

As a part of the engagement process in the Talent Management Model, sites strive to foster open communication at all levels in a positive work environment. As a result, Associates have become more and more involved in the Lean journey and have been critical to our success. All sites hold some form of Town Hall meeting to keep personnel informed. General topics include: Safety, Quality, Operations, State of the Business, etc.

Problem Solving Processes and Team Roles

Accuride leverages several forms of problem solving processes depending on the complexity and/or nature of the issue:

1. **4SQ.** A 4SQ is an action item tracking tool that is StdW across Accuride. It is color coded for easy identification of status and denotes timing and owners. The project owners maintain living 4SQs.
2. **Fishbone/Ishikawa Analysis.** This tool is used to categorize potential causes for corrective action.
3. **5-Why.** 5-Why's are conducted with cross-functional teams using a drill deep methodology to identify true root causes. 5-Why's can be used to drive long-term corrective actions within the 8D process.
4. **8D CA/PA.** 8Ds are used for external and internal items that need an in-depth analysis. The 8D CA/PA process meets TS/IATF 16949 standards for the Automotive Industry. The 8D form includes basic categories for problem solving, such as corrective actions, verification of corrective actions, preventive actions, interaction with Control Plan (CP) and Process Failure

Mode and Effects Analysis (PFMEA) and a 5-Why analysis. This process drives thorough and effective root cause identification and permanent corrective actions that are subsequently verified for effectiveness. It is used to address major customer concerns, internal or external audit findings or other issues where an in depth analysis is deemed necessary. An 8D CA/PA log is maintained at the corporate level for cross-comparison within ACW for both Lessons Learned and Prevention.

Approaches to Reward and Recognition

Accuride associates treat one another with respect. Ethics and integrity are built into ACW's Strategic Vision Part of this process in recognizing our associate's achievements and demonstrating true appreciation. Accuride recognizes length of service for significant anniversaries. Multiple types of awards are given, such as gift cards and watches.

Data and Evidence of Employee Morale

It has been demonstrated that the more personnel are engaged in the operations, the better the morale of the overall team. Turnover and attrition has been significantly reduced year-over-year. This improvement suggests that our associates are more stable and satisfied with Accuride. **Figure 7.7** shows a set of survey responses from one of the AME award winning sites.

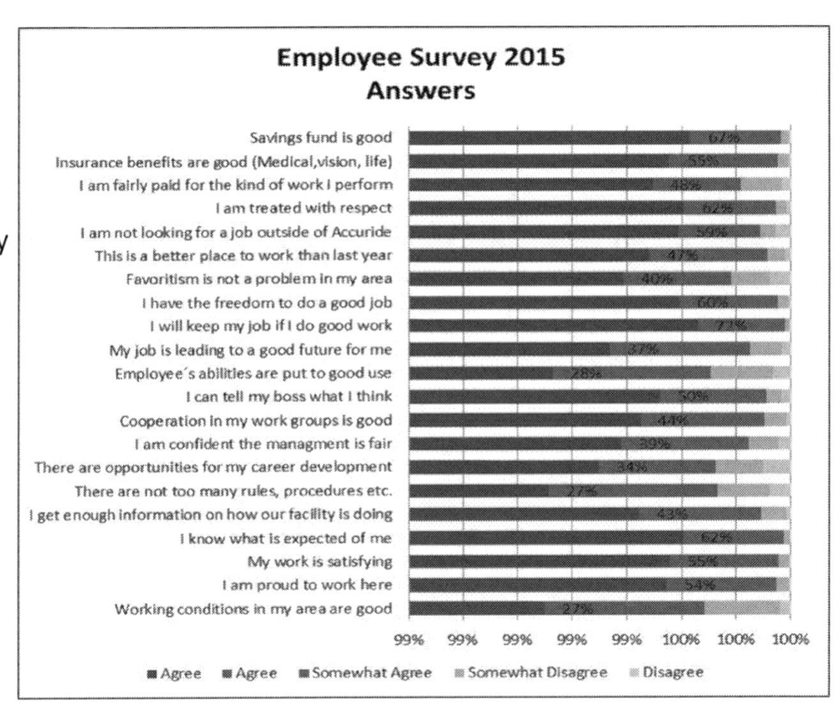

Figure 7.7

Section II: Manufacturing and Business Operations

Section 2.1: Manufacturing Operations

Accuride's Lean program leverages more than 20 standardized forms of VOS, which are designed to provide both transparency and process standardization to both manufacturing and business operations. The VOS is incorporated into Accuride's corporate QLMS in the CI Section. An initial key tool is the VSM. There are four forms of VSMs: C-VSM, F-VSM, T-VSM and E-VSM. The F-VSM is always expressed as three months into the future. An E-VSM depicts product and/or process flow outside of one facility and includes sister sites, warehouses, customers and suppliers. The C-VSM is initially created on key processes with a tremendous amount of detail. Takt time is calculated and a resulting timeline for the product/process is drawn with a focus on reviewing how to reduce and eliminate both defects as well as lead time. Kaizen bursts are identified, and a detailed 4SQ is developed to manage process improvement. There have been multiple iterations of the C-VSM/F-VSM at all the sites. T-VSMs are conducted on transactional processes typically within the business. Accuride understands that he who delivers the product the fastest (with quality and competitive costing) wins (Quote from Rick Harris).

Identifying and Eliminating Waste (Muda)

Refer back to **Figure 4.22** to see how Accuride cross-correlates some of its key QLMS tools with the types of losses typically encountered in its processes.

Overproduction: Accuride has machine lines/assembly lines that serve as Pacemakers for the entire Pull System. Product is pulled from a supermarket to these lines. Product is fed to the Pacemaker directly via FIFO from the WIP supermarkets.

To aggressively minimize overproduction, Accuride has deployed a PFEP. The PFEP tool analyzes several factors, such as the forecast, OEE, MTBF, Supplier Reliability, Shipping Reliability, etc., and then, based on Takt time, it determines the optimal market size. Next, the best-fit supermarket inventory and the production run size is determined and the changeover schedule is developed. Kanban cards are then used to flow the product from raw material storage through the pacemaker processes to finished goods.

Raw Material and Packaging Supply inventories are also managed by a PFEP and delivered at fixed intervals. 100 percent of Accuride's Raw Material, Packaging Materials and Finished Goods are managed by PFEP. As a result of the dedicated execution of the PFEP and other supporting lean processes, there was a substantial reduction in Excess and Obsolete inventory (E&O). DIOH has steadily declined. The decrease in DIOH has reduced working capital since the on-set of our Lean journey.

Waiting has also been positively impacted at Accuride. Coupled with the PFEP process, strategic kanbans were incorporated across the stream. A leading indicator metric called *%Pull To Plan* was implemented to measure the team's effectiveness of properly pulling to the kanban cards. Kanbans have been implemented both within and external to Accuride. In addition to kanbans, a tugger system was established to ensure that our associates have the necessary materials on hand to do the job.

Transportation Time, too, has been reduced. While counterintuitive, product can be hidden "in plain sight" when there is too much rolling stock, such as fork lifts. With internal WIP and FG quantities reduced by such significant amounts, the need for almost 34 percent of the rolling stock, such as forklifts, went away. What was no longer needed to be produced no longer needed equipment to move it. Accuride modified the receiving docks to transform the flow of the plant. This transformation reduced travel distance significantly.

Excess Process Time has been reduced with some of the above as well as with some focused quality tools, specifically Attribute Agreement Analysis. AAA is a form of measurement system analysis for attributes. It is used to calibrate the people performing the evaluation. Accuride uses AAA to ensure that associates are making the right disposition calls on products, thereby minimizing rework and causing lost capacity. An extensive multi-plant AAA was performed to calibrate all sites that make similar products for surface finish requirements and defects.

Excess Inventory has essentially been reduced at Accuride since the start of its Lean journey. We make different assumptions for Runners (high volume items), Repeaters (regular items), and Strangers (make-to-order, low volume items). This is managed through the PFEP and TTR. TTR is a Heijunka model that organizes the production order of the products in the process by Runner, Repeater and Stranger as well as in "set-up" order for degree of difficulty. This model is managed by kanban cards that determine how much of each type of product will be produced. When that limit is achieved, then the line is changed over. Effective TPM is a necessary system for this process to be effective.

Excess Motion is largely targeted within Accuride by StdW, which is an output of conducting Operator and Machine Balance studies. Operator Balance charts are created for production and conveyance processes. These charts are being used to rebalance work opportunities and to optimize staffing. This process has been implemented to achieve a significant improvement in productivity since the start of our Lean journey. Excess personnel were redeployed. It should be noted that the majority of redeployments have been implemented without the use of non-voluntary layoffs. Unfortunately, this cannot be said for volume adjustments, which our markets are heavily subject to.

Defects are attacked with a vengeance at Accuride. Our product is safety based and it is critical that effective prevention and detection systems are in place. Accuride sites track all KPIVs "from wall to wall" for all features deemed as CTQ. These may be speeds, feeds, temperatures, dimensions, attribute features, etc. These items are also identified in the Control Plan per AIAG systems criteria. A master list of KPIVs by site is monitored on a monthly basis as to whether or not it meets a Cpk of 1.33 or greater. Processes that do not meet a 1.33 are generally 100 percent inspected for conformance. Accuride monitors the percentage of KPIVs that are capable. The overall result is reported to the EC on a quarterly basis. Within the month, SPC is used to ensure that variable-based KPIVs can be sustained. As a result of this focus, scrap has dropped year-over-year during the Lean journey. The KPIV metric is used as a leading indicator for scrap and COPE. First year warranty issues have been significantly reduced since the start of the Lean journey.

Unused Creativity is tapped through a variety of processes. Accuride uses a formalized Gemba walk to manage the facility with visual systems. During this walk, associates are interacted with as a means to capture creative input. There are other methods used to ensure associates' suggestions and ideas are captured, including safety suggestions programs and monthly Town Hall meetings.

Identifying and Eliminating Unevenness, Fluctuation and Variation (Mura)

Addressing unevenness is a large area of focus at Accuride. As previously outlined in **Figure 4.22**, many of the systems provide multiple benefits for mitigating both Muda and Mura. Most notably, these are StdW, TTR (Heijunka), kanban and Value Stream Mapping. Due to the safety nature of Accuride's product, quality and production associates are empowered to stop the line should a concern arise. It is cited on the empowerment matrix. Effective implementation of PFEP has significantly reduced exposure since there is much less product in the pipeline. Error Proofing is managed in the 5-Why and 8D CA/PA processes. The Accuride team has used these tools for proactive and reactive identification of root cause and elimination of recurring of issues. The team has leveraged the support of several Master Black Belts to minimize waste and ultimately reduce concerned inventory. One KPI that acts as a leading indicator to OTD is %Pull To Plan. It monitors the accurate and correct pull to the kanban signals.

Another part of managing unevenness at Accuride is Preventive Maintenance (PM). A comprehensive visual system is installed via a series of PM boards, which indicate the status of all key assets in the facility. All PM routines are issued by and closed in Plex creating a closed loop. Visual PM boards are then updated with the latest information. MTBF is tracked and analyzed for all key assets and is reviewed regularly for opportunities to improve. MTBF is included in the quarterly KPI review.

Reduce Burden on People and Equipment (Muri)

Accuride has a strong focus on safety for both its people and its products. Managing burden on people and equipment is key to Accuride's success. When overburdening of people or equipment occurs, it can cause fatigue-based safety incidents, errors in judgment, quality and delivery errors, damage to capital equipment and a variety of other negative effects. Many of the systems provide additional benefits for mitigating Muda and/or Mura along with Muri. These are most notably TTR, kanban with continuous flow and Plant Layout to manage material flow.

5S is a standardized process across Accuride. Accuride embraces this system, which includes a series of proactive audits to ensure sustainment. Designated clean up time is built into the production schedule to ensure that the organization of the facility is incorporated into the overall Lean system. Area audits are performed every month, including production lines, labs, maintenance shops and different warehouses. This information is located in the Glass Wall area.

Outbound Shipping Boards: While control of shipping and receiving largely impacts Mura (unevenness), Accuride associates felt the stress more deeply when trucks were backed up for loading or unloading. Errors could be made, accidents or handling damage could occur, etc.

Accuride utilizes a simplified system for scheduling outbound loads.

Changeover Improvements: The Accuride teams have put considerable focus on the changeover process in the production process. Accuride's ability to perform safe, time efficient changeovers has been a key component to its inventory reduction and enabled the team to get around the changeover wheel with greater velocity. A Set-Up Reduction (SUR) methodology was developed to focus on reducing time. The steps incorporate, separation of internal and external tasks, shifting internal tasks to external, streamlining external tasks, locating parallel tasks, streamlining internal tasks and practicing the new method.

Automation Improvements are also key to minimizing burden on people. Accuride has several large scale robots in place in the facility for different processes.

Visual Operating Systems (VOS): As discussed throughout this book, Accuride uses Gemba tours based on observation of the VOS in the plant. In short, as our team walks around, the plant should "talk" to us. These visual systems are the backbone to Accuride's Lean system.

Section 2.2: Business Operations

Accuride recognizes that Lean is not just an "operations" tool. As previously shown, **Figure 7.8**, Accuride's Lean Path Model, three types of streams are critical to business flow: Operations, Transactional and External Partners. All on-site staff have undergone intensive internal Lean training while evaluating these various flows.

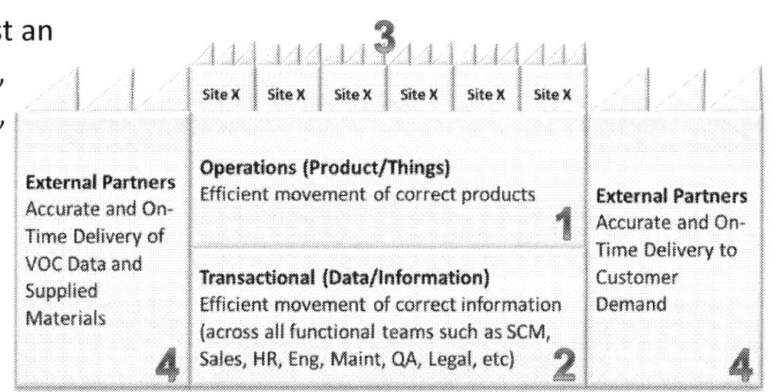

Figure 7.8

Identifying and Eliminating Waste (Muda)

The Accuride team has implemented transactional improvement projects ranging from the Financial Close Process to timely Issuance of Order Entries. These improvements have been realized by use of a T-VSM. Across Accuride, each site has taken the lead on one to three core processes. As each site develops their T-VSM and supporting StdW, they share it across the other locations. Accuride calls this replication. This has had several benefits:

1. It enables corporate-wide StdW for shared transactional processes.
2. It reduces the lead time for analysis and execution.
3. It acts as Lessons Learned during the implementation of the improvement(s).

This methodology has been well received across the business. At the site level, this method of replication typically addresses **Overproduction, Waiting, Transportation Time, Excess Process Time, Excess Inventory, Defects and Unused Employee Creativity**.

Identifying and Eliminating Unevenness, Fluctuation and Variation (Mura)

Unevenness is largely addressed through the replication model described previously. Conducting T-VSMs enables Standard Work, Mistake Proofing and Enterprise Flow via the management of information flow and information sharing. Another key process that Accuride has implemented is the establishment of pull programs and strategic partnerships with its key MRO suppliers, a sample is shown in **Figure 7.9**.

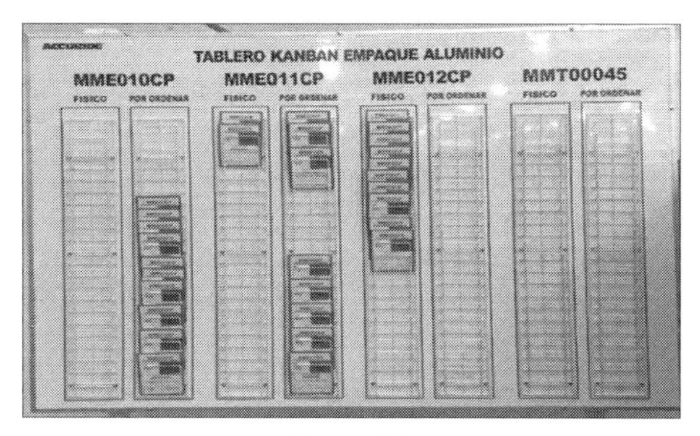

Figure 7.9

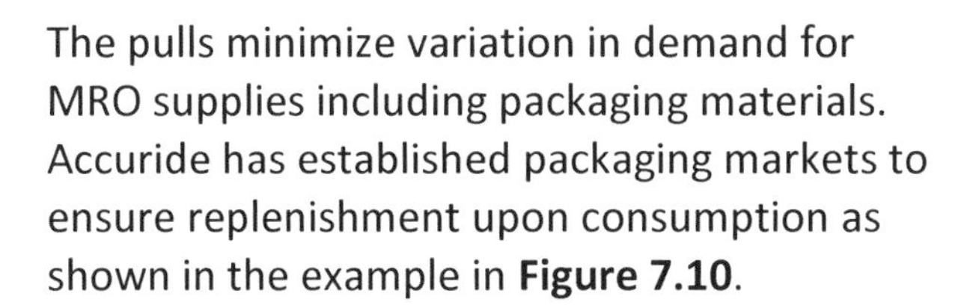

The pulls minimize variation in demand for MRO supplies including packaging materials. Accuride has established packaging markets to ensure replenishment upon consumption as shown in the example in **Figure 7.10**.

Figure 7.10

These level pulls have allowed Accuride's suppliers to share in the benefits of a Lean system. Accuride has also implemented tracking of MRO materials to be continually rationalized if the inventory is not appropriate for current conditions.

Reduce Burden on People and Equipment (Muri)

As described above, the Level Scheduling T-VSM reduced burden on our associates. Our supply chain team reduced the total process time, one of the biggest impacts on this process was the implementation of the pull system schedule and kanban boards to optimize the process.

Section III: Extended Value Stream Management

Section 3.1: Product Development

Process Innovations Used to Meet Customer Expectations

Customers have continued to communicate the need for lighter weight and improved corrosion resistant steel and aluminum wheels and cast iron drums. To better service the customers, Accuride has implemented weight savings programs and increased finish quality offerings. The Accu-Lite® steel and aluminum products have reduced the weight of the finished product, and therefore improved fuel efficiency of the trucks. New product is in development for an aluminum-based drum. These improvements have helped the global environment by supporting the emission reduction requirements of trucks. Accuride has also launched new coatings to address the fleets' needs for better corrosion resistant wheels. These are reviewed in other chapters.

Develop Workforce Understanding of Customer Expectations

Accuride is actively connected with its direct OEM customers and distributors as well as the end-user customers; the trucking fleets. Throughout the year, various forms of information are collected and analyzed to develop the VOC. There are events each year to evaluate VOC feedback for key customer groups, such as OEMs and Aftermarket.

Accuride engages in a detailed commercialization process (AccuLaunch™) where a cross-functional team is involved to evaluate the customer's expectations. AccuLaunch™ starts with a detailed business case approval process and then is followed by several "gate" reviews. Each program is managed through this review structure. Across the process, several key documents are created. These include a Design Failure Mode and Effects Analysis (DFMEA), a PFMEA, CP, relevant operator Work Instructions (WIs), Blueprints and Engineering Specification (ES) documents and several others.

The model used by Accuride closely follows the Advanced Product Quality Planning (APQP) guidelines set forth by the AIAG Guidelines and meets the requirements set forth in ISO/TS 16949 (IATF 16949). The function of the PFMEA and CP most specifically is to translate customer requirements and CTQ features into the production facility. Personnel training is conducted with these resulting documents so that CTQ features are understood during the product transformation process.

Innovative Processes Used to Decrease Cost and Increase Value

Accuride's Wheels Product Engineering is located in Henderson, Kentucky. Design and testing are conducted at this location. The engineering team leverages an effective commercialization process (AccuLaunch™) where OEM account managers and sales engineers collaborate with customers to identify their application requirements for new wheel designs. If a customer requires prototype wheels for analysis, Accuride has a fully dedicated R&D department capable of producing sample wheels to meet any customer's specifications and requirements. Accuride's state of the art testing facility is also available for conducting both physical and metallurgical testing per American Society for Testing and Materials (ASTM), Society of Automotive Engineers (SAE), Technischer Überwachungsverein (TUV) and Association of European Union Wheel Manufacturers (EUWA) standards. ACW's LBF tester is the only one in North America large enough to test heavy truck wheels.

Leveraging New Product Development to Minimize Total Cost

Accuride utilizes state of the art Finite Element Analysis (FEA) and metal forming software to analyze wheel designs and manufacturing processes. Both of these tools enable Accuride to design light weight wheels in addition to optimizing the material flow in order to provide an optimal manufacturing cost structure. Accuride's product engineers also conduct design review meetings with the manufacturing facilities to seek input on their designs. These meetings include focus on tracking project timing and spending to meet customers' requirements.

Accuride's Approach to Benchmarking

Accuride's product and field engineers are actively involved in the Tire & Rim Association (TRA), SAE and other trade organizations throughout the commercial vehicle industry. By either leading or actively participating in these organizations, it allows Accuride to monitor and develop wheel designs that meet the ever changing environment of the commercial vehicle industry. Accuride also works closely with its material suppliers to monitor and develop new products for wheel applications. This involvement with Accuride's material suppliers has allowed Accuride to reduce the weight of a steel wheel and an aluminum wheel, thus yielding significant fuel savings to the commercial vehicle industry.

Accuride's associates regularly visit peer site locations during the quarterly Lean reviews. As an AME member, Accuride personnel also conduct benchmark reviews at external sites.

Accuride's Focus on Variety Reduction, Commonality and Modularity

Accuride has been designing and manufacturing steel wheels for over 100 years and aluminum wheels for over 40 years. Any time a new wheel design is requested by a customer, an extensive effort is made to see if an existing wheel or component can be utilized to meet the customers' requirements. Accuride reviews its existing product portfolio annually and determines if lower volume wheel designs can be consolidated with other existing wheel applications or eliminated from the portfolio. Accuride can provide customers a complete wheel-end system. Commonality must be maintained between component families. **Figure 7.11** shows the complete wheel end system.

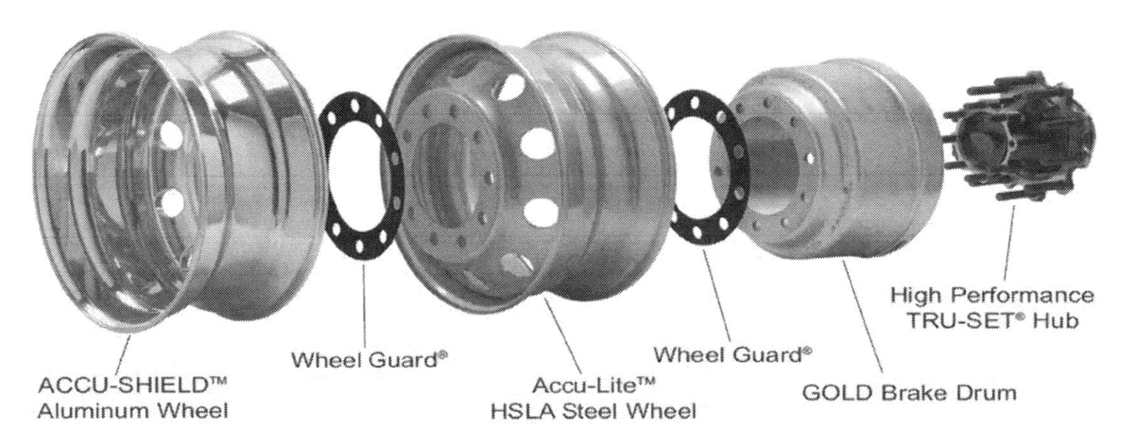

Figure 7.11

Section 3.2: Supplier Development

Supplier Partnering to Minimize Total Cost of Value Stream

Accuride recognizes that Lean is not just an "internal business" tool. As previously shown in **Figure 7.8**, three streams are critical to Accuride's business flow: Operations, Transactional and External Partners. This includes Accuride's supply chain. Suppliers are business partners and Accuride works with them closely. Key Accuride suppliers participate in a pull system guided by the PFEP process. The PFEP is used to trigger releases or pulls upon consumption for supplier deliveries. Supplier kanban boards enable us to monitor and ensure raw materials are within the established levels and avoid shortages. Across the site, minimum/maximum standards are established so that third party suppliers with Vendor Managed Inventories (VMI) can plan appropriately. This also significantly reduces the risk of over or under ordering. From a general flow perspective:

1. Kanban card systems are used to ensure replenishment occurs in real time.
2. Tugger routes are used to replenish supplies to identified line locations.
3. Communication between the supplier and Accuride is enhanced.
4. Lead times are shortened.
5. Expedited orders become less and less frequent.

Suppliers are included in Accuride's C-VSM, F-VSM and E-VSM activities. This ensures that the receiving processes within the value chain are included in Accuride's analysis.

Supplier Certification Processes

When evaluating a potential supplier, Accuride references Corporate-level procedures. These procedures outline the requirements for suppliers, the process for evaluating them and the process for continuing to monitor them through key measureables. Requirements for the suppliers could include ISO/TS 16949 (IATF 16949) and ISO 17025. A key document when certifying an Accuride supplier is the Approved Supplier List (ASL). The ASL includes ISO or TS certified suppliers that have been trialed and tested through appropriate engineering tests and/or through PPAP. The ASL also functions as a tracking system to identify an Approved Supplier's Certification status.

Suppliers for key materials that can affect CTQ features often submit an in-depth PPAP package. This structured document process meets ISO/TS 16949 and AIAG systems criteria. Due to the safety nature of Accuride's products, Accuride personnel will often coach and assist suppliers in the development of these documents to ensure that effective and efficient systems are in place.

Supplier Focus for Continual Improvement & Business Results

Accuride maintains a publicly available Supplier Quality Manual through its website. **Figure 7.12** shows the Accuride webpage for supplier standards and processes. This outlines Accuride's expectations of suppliers, giving them the tools to help Accuride meet and exceed business plan objectives.

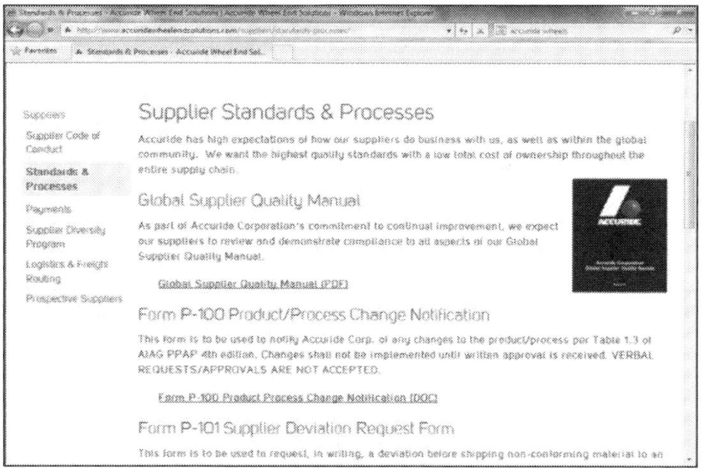

Figure 7.12

Effective partnering with suppliers is key to obtaining high quality products and services. Accuride and its suppliers are aligned to the same goal: satisfy the end user. Examples of these partnerships includes vendor management, consignment programs, consumption kanbans, annual guaranteed cost savings with key suppliers, inside technical support, etc.

Processes to Achieve Perfection in Product and Supplier Management

Accuride has documented procedures that outline measureables for key approved suppliers, which are reported on a regular basis. These measureables are then used to populate a scorecard. These scorecards include various types of information such as: Corrective Actions, Supplier Internal Performance, Key Measureable Data, Potential Cost Savings, etc. This information is used in the Semi-Annual Supplier Review.

Innovative Processes to Improve Market Service and Logistics

Accuride's E-VSM captures our multi-national value chain to continually improve upon flow and reduce working capital. Each site enjoys excellent support and relations from its sister facilities. There is strong inter-company support. If there are local issues relevant to Accuride's sister sites, the local on-site team will provide local support and evaluate product concerns. They will visit local customers or suppliers on behalf of these sites to shorten response lead times.

Sales and marketing has a close relationship with customers allowing Accuride to monitor and respond to customer needs in real-time. Decisions to increase aluminum capacity in North America, introduce an industry leading coating technology, provide improved packaging solutions to customers and establish Economic Order Quantities to improve plant performance by reducing complexity, are all results of close interactions between Engineering, Sales/Marketing and Customer Service.

Key Takeaways

1. Jim Womack & Daniel Jones have identified five principles to Lean thinking: specify value, identify value stream, make value flow, let customer pull and pursue perfection.
2. One needs to distinguish constraint from bottlenecks and keep working to address these constraints. However, for the purpose of scheduling, only the pacemaker (may or may not be a constraint) needs to be provided the schedule since after the pacemaker the product/service is supposed to flow to the customer.
3. In order to pursue perfection, one should focus on adequacy, effectiveness and efficiency of the process inputs.
4. Accuride Lean system is monitored via monthly Lean council calls by the Lean council and quarterly Lean review by the executive council.
5. Accuride Lean system follows X-Matrix (Hoshin Kanri) and is implemented by the motivated teams with excellent problem solving skills.
6. Accuride Lean system extends to transactional processes across multiple functions as well as beyond the organization to suppliers, outside service providers and customers.

Chapter Eight: Building Partnerships

Supply chain management is well positioned to initiate relationships and to evolve supplier and customer partnerships with the full visibility and engagement of senior leadership. This has become an asset to Accuride. Because of increased globalization, increased interaction between internal and external disciplines and the need for advanced technology, Accuride's supply chain team has assumed the leadership role in "building partnerships" and has transformed the global supply chain. This exponential transformation within Accuride's global business environment, measuring organizational performance, and has introduced the requirement of understanding and developing a network of collaboration and partnerships.

Accuride's supply chain originally consisted of the traditional purchasing and inventory control personnel, working independent of the business units at the corporate level. Today, the team spans seven different disciplines, creating full visibility from order entry to delivery of the product to customer's. The creation of such a structure is no simple task, but when executed correctly adds value to multiple aspects of the enterprise. The supply chain team started with the integration of global purchasing, inventory management and planning, logistics and the addition of supplier quality and development. As time went on, it was evident that full visibility was not there. Though performance metrics were in place for cost savings, on-time delivery and supplier performance, it was clear that customer expectations were still not being met. In the absence of integrated disciplines, emerging challenges were not being addressed. To truly meet the demands of Accuride's original "Fix and Grow" strategy, its supply chain needed a more robust structure. **Figure 8.1** shows a T-VSM flow of the evolution of the MRO process from fix to streamlined control.

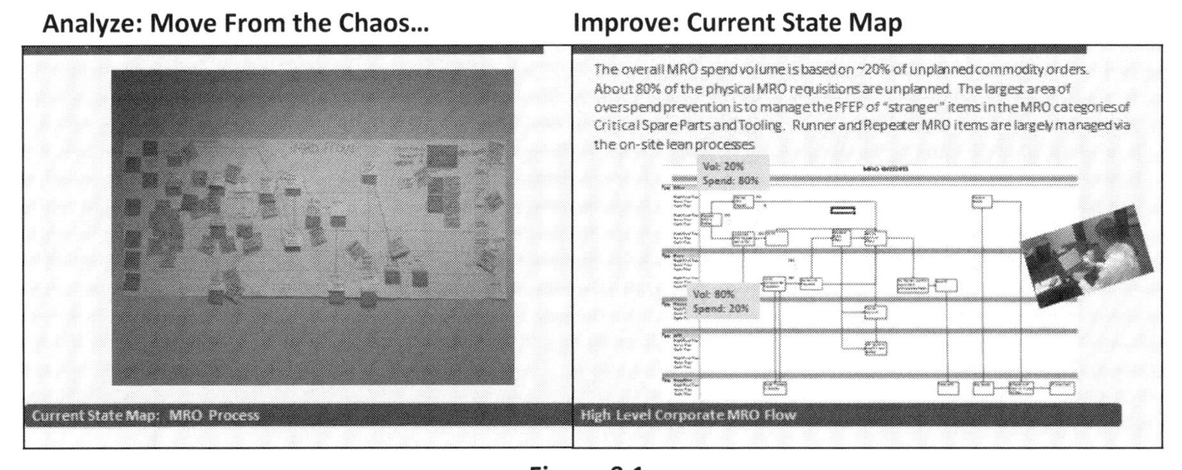

Figure 8.1

Packaging, internal sales and distribution were added to the supply chain function, hence, creating full visibility from customer order entry through distribution of the final product to customers. The performance of these disciplines became critical to understanding what was needed to "fix" the organization, and realize the growth sought by the executive leadership team. The interface with suppliers, as well as customers, expanded, creating a stronger correlation between supply chain maturity, performance and the level of the relationships internally and externally. The strategy was to influence relationships and create partnerships, which would contribute to the level of knowledge within the company and bandwidth of experience, through the engagement of leadership collaboration.

Having a clear definition is the starting point to drive results and the development of supplier and customer relationships. What does the partnership need to look like to achieve the end results? The critical success factor is a need and desire from all parties to collaborate. For this baseline to be established, the need for trust and full visibility must be understood. The direction and commitment comes from the CEO and the executive leadership team.

The greatest challenge was full alignment of operations, functional teams and the prospective "partner" objectives. The need for constant communication, trust with sharing corporate information and full collaboration was entangled with the ability to compete. Multiple contributing factors aid and assist in the creation of solid partnerships. Within Accuride, the consistency of leadership had been previously criticized by both suppliers and customers. The CEO, was setting the pace and changing the perception by establishing a solid executive committee team with expertise in their global functions. The consistency of leadership, coupled with a strong vision brought in by all, made the difference. As cross-functional accountability was established, the focus shifted to building strong partnerships, orchestrating goals and objectives and ultimately aligning with corporate strategies.

As part of the goal to change the perception of Accuride, leadership worked on specific data and facts to be shared not only with Accuride associates, but also, just as importantly, with Accuride's partners. Information sharing with strategic suppliers and customers was established, which included sales forecasts, delivery performance, commercial market demand levels, capacity utilization, operational Lean performance and corporate short and long-term objectives. With this type of relevant information sharing, suppliers and customers saw that the team had the facts and data as well as the willingness to share the information and take full accountability for the results. Over time, strategic partners saw synchronization in decision making. Team members were being held accountable and the ability to make joint decisions evolved.

Suppliers and customers began to weigh in on the balance of power, resource needs, classification of objectives and strategy development from a long-term perspective. Collaboration in relationship building allowed for incentive alignment for Accuride as well as its suppliers and customers. Information sharing and visibility into Accuride's executive team beliefs and strategies became strong predictors of whether or not the partnership would be successful. The sharing of partner strategic objectives, competencies and vision was achieved through strategic executive leadership collaboration on a routine basis. The partnerships were being sought for the long-term, not for the traditional contract negotiation. These partnerships were initiated and facilitated by executive team members.

The steps required to arrive at positive, influential results became more evident as information was being shared. Open communication was a major contributor, use of multiple levels of information, participant's roles and freedom to explore multiple topics became the norm in partner strategic meetings. Top executives from suppliers and customers became engaged and began to engage in Accuride's Lean journey, performance and objectives. As results in performance became transparent, the focus shifted from tactical communication to strategic development. The "involvement" of multiple disciplines, internally and externally, became the norm. Overall WIP and FG inventories were able to be reduced, in part, with the planning and reliance on supplier performance as shown in **Figure 8.2** and **Figure 8.3**.

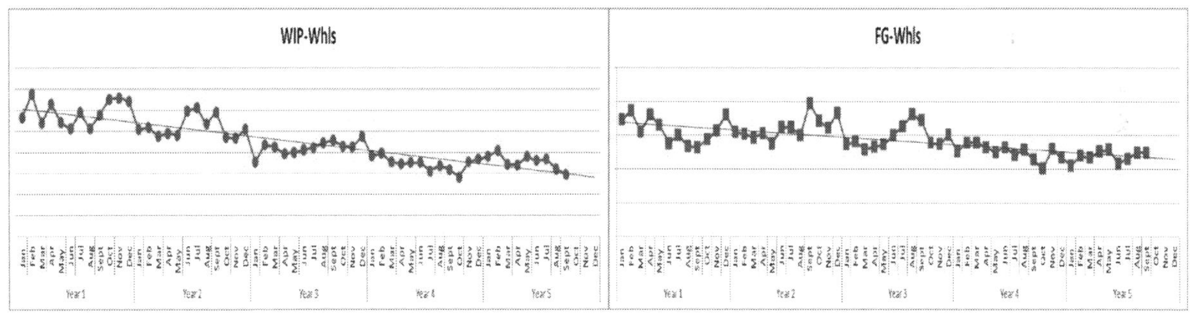

Figure 8.2 **Figure 8.3**

In order to evolve from a global supply chain philosophy, into a mindset of collaborative partnership development, Accuride had to go deeper. The potential value was evident, but "involvement" had to work within the functional areas, creating collaboration at all levels of the organization.

With specific performance metrics established and communicated, the collaboration had to be stressed across multiple dimensions. These dimensions had to complement each other while contributing value to both Accuride and its partners. Performance can improve if all team members understand and support the involvement at multiple layers within each organization. An example is on-time delivery performance and internal sales. Key performance was on-time delivery to our Customer Required Date (CRD) and Mutually Agreed to Delivery Date (MADD). Specifics were being tracked; such as customer drop-in orders not within Accuride stated lead times, plant issues, transportation carrier issues and Accuride performance. These all affected Accuride's level of Past Due. See **Figure 8.4** for recent Past Due trends.

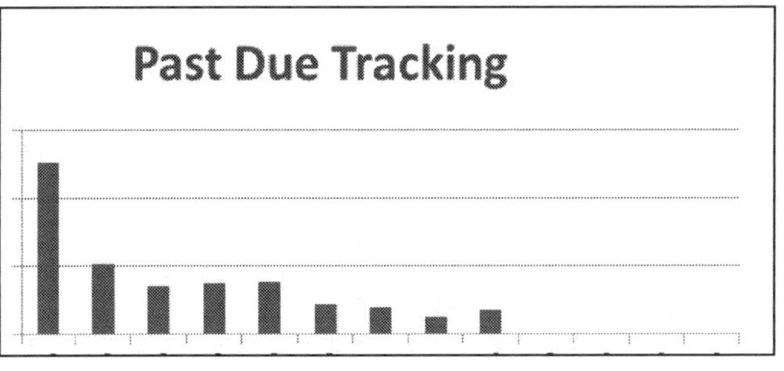

Figure 8.3

With full visibility, Accuride's internal sales needed the involvement of the sales team, supply chain team, customer planners and ultimately the senior leadership at the customer to align the data associated with performance. Without the involvement of customer functional leadership, the information would not have been shared, and the transparency required for improvement would not have been realized.

Several of Accuride's customers are on parallel Lean journeys and have proactively engaged in pooling their resources for mutual benefit. One major customer collaborated with Accuride's Supply Chain and Lean teams to reduce their on-site inventory footprint by more than 50% while sustaining the same level of pull (**Figure 8.4**). While these value added processes do not usually show up on the customer's scorecard of Quality, Cost and Delivery from a component perspective, they do add intrinsic value.

Figure 8.4

Supplier collaboration is no different than customer collaboration. Though perception across the industry often reflects that "full visibility" could be giving away trade secrets, Accuride has proven that is not the case. The practice of leadership collaboration and supplier involvement has influenced the strategic supplier partnerships. The level and quality of information shared at all levels of the organization has enabled Accuride to leverage the competitiveness of its entire business network, as opposed to simply competing as an isolated company.

Contributing factors again relate to the level of information being shared as well as the quality of information being shared. The communication to our business partners across multiple functions became a building block and is developing trust and knowledge with Accuride suppliers. The routine of strategic supplier meetings covered Accuride needs, performance and strategic goals. The suppliers shared their industry insights. This level of knowledge sharing contributed to the achievement of an integrated supply chain, from supplier to customer. Effectiveness is now measured by the introduction of advanced technology, exclusivity agreements and long-term strategic alignment contributing to the elimination of uncertainty in the supply chain and improved supply chain performance. See Chapter 9 for specific examples of this success.

Other examples are described below to illustrate the needs, requirements and practices that contribute to Accuride's success with leadership collaboration across multiple functions, which significantly increase Accuride's competitive advantage along its Lean journey.

Raw material is a major, significant spend within Accuride. The exposure is great not only from a dollar perspective but from a manufacturing perspective. The ability to have raw material suppliers as strategic partners influences not only the potential elimination of supply uncertainty but leads to a competitive advantage through product quality, delivery and technology. Accuride's executive leadership engaged certain leaders at specific raw material providers to establish greater collaboration. The executive team sought those suppliers that had similar values and beliefs as well as those who sought global growth were committed to full transparency. A valued long-term supplier came to the surface quickly. Aligned with Accuride's "Fix and Grow" strategy, this strategic supplier's executive leadership stepped up and engaged in open communication. This level of communication included their strategic objectives, performance criteria and perception of Accuride's performance, including team member performance, current state of the existing relationship and perception of the company within the industry.

It was hard to hear some of the feedback that the suppliers were providing. The leadership team had to be respectful and reflective of the information received and leverage it in the spirit in which it was provided; to positively evolve the relationship. Consistent leadership meetings were held. Alignment was considered to ensure that both Accuride and the supplier's needs were being met. Multiple disciplines participated including research and development, product engineering, sales, supply chain and executive team members. Quarterly reviews were established. The quarterly reviews grew from valuable information sharing meetings into global growth strategy development with technology advancement discussions and collaborative sessions regarding how to best move to the next level. Collaboration was structured around advanced technologies, competitive advantages and long term relationships. Outcomes included the introduction of new technologies with exclusivity agreements and a competitive advantage position for both the supplier and Accuride. A solid partnership was built on involvement and contribution from both sides.

To improve performance and accountability, Accuride initiated the joint use of tools for tracking performance and measuring success between the company and its supplier partners. Driving supply chain excellence must include improving and creating value in performance. Both the customer and the supplier must utilize a shared set of tools. Accuride made the strategic decision to utilize two specific tools to evolve to the next level of performance. Customer Relationship Management (CRM) and Transportation Management System (TMS) were the two primary tools implemented.

Sales Force automation tools, in particular CRM, provide an organization with a mechanism to consolidate key customer information, identify and qualify new business opportunities, manage opportunity pipelines and enhance communication to ultimately drive top line revenue growth (**Figure 8.5**). Accuride's complex selling environment provides a unique challenge; multiple product lines, multiple sales channels and a "many to many" sales relationship to the ultimate end customer, the commercial fleet.

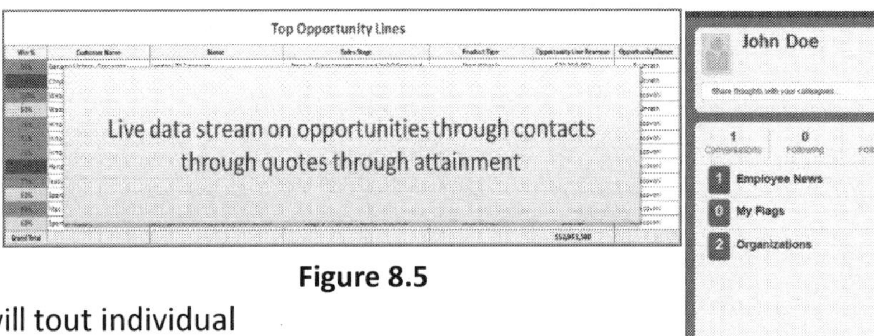

Figure 8.5

Each CRM module provider will tout individual features and benefits. The key differentiators in Accuride's selection process, and the final sourcing decision, were the opportunity/pipeline management, Outlook integration, mobility and general ease of use. Today we have a tool in place that provides:

- Custom reports; utilized for team meetings: Sales Pipeline, Fleet Demographics, OEM summaries.
- OEM summaries auto-generated and e-mailed (weekly).
- Ability to readily transfer Account ownership: Opportunities, contacts and account history remain.
- Mass E-mail notification to affected customers.
- Support short-term coverage gaps.
- Mobility features: "Around Me," voice access for meeting notes.
- Oracle Social Network (OSN): Repository for key information (OEM option codes, OES DPM contacts, training guides).

We continue to review the effectiveness and assimilation of the tool with our front line selling team. Several "enhancements" have been driven by our mid-term evaluation:

- Variance reporting: Side-by-side snapshot of current vs. prior week/month.

- Clear identification of conquest vs. retention of current business.
- Accuracy of remaining opportunity value.
- Reviewing configuration parameters: Team feedback indicates desire to improve speed.

Oracle Sales Cloud was launched in October, 2014. Accuride's team approach to selling (with establishing baselines) was an "ours to lose" vs. a true "conquest" of sales. This presents a challenge to identify/track individual performance metrics within the CRM. There are many success stories that can be highlighted. One example was with a key customer's pull-through demand for a new product line, Steel Armor™. Year-over-year for 2014:2015 was 116 percent and year-over-year for 2015:2016 was 142 percent. This represented a true team effort between engineering (technology) and sales.

The team also implemented and executed Transportation Management Systems outlined in **Figure 8.6**.

TMS Transportation Management Systems

- **Streamlined process** for all locations: Leverages corporate pricing, all shipping under one process
- **Carrier Procurement**: Negotiation of carrier pricing, contracting carriers, carrier rate loads, network analysis
- **Logistics Operations**: Leveraging market rates, optimization of Accuride customer/vendor network
- **Freight Audit and Payment Services**: Verification of rates
- **Elimination of Decision Making; Let the System Drive the Process**: Error proofs the carrier selection criteria per lane, per load
- **Efficiency**; Eliminated Touch Points for Accuride Personnel: People can focus on internal improvements
- **Freight Spend Control**: Real time data analysis

Figure 8.6

In summary, performance metrics were created and implemented, creating not just reduced cost and improved service, but positively impacting shareholder value. Strategy was developed, supply chain levers were added and Accuride forged forward in the development of strategic partnerships with full visibility throughout its supply chain. Statistically, performance was enhanced, knowledge was expanded and the value of the established strategic partnerships began to positively influence shareholder value.

Performance was indeed improved, the development of full visibility, open communication and strategic partnerships continue to contribute to Accuride's success. Though we are still on our Lean journey, the performance speaks for itself:

- OTD went from 54 percent to 99.8 percent.
- Lead times were reduced by 18-87 percent with an overall average exceeding 55 percent.
- Supplier OTD improved by 35 percent from baseline.
- Inventory reduction of 28 percent year-over-year.
- Transportation and audit cost reduced by 23 percent

A level of appreciation exists among our global strategic partners. We look forward to our future alignment and long-term partnerships.

Key Takeaways

1. Lean management must include external partners, such as supply chain and customers. Benefits include bottom line performance improvements as well as intrinsic partnership benefits that competitors cannot provide.
2. Leading indicator KPIs are just as effective with external partners as they are with other parts of the business.
3. There are key systems, which can be incorporated into the overall process stream to better optimize supply chain effectiveness and efficiency, such as Customer Relationship Management (CRM) software and Transportation Management System (TMS) tools.

Chapter Nine: Show Me the Money… Or Not

Lean is an effective set of systems designed to optimize processes and save money. That's why businesses find it such an attractive concept. However, when they get right down into it, it can be challenging to quantify where those savings are coming from. It's not a rapid panacea. To be the best in the world and to really leverage Lean systems, the entire organization needs to be engaged to realize the benefits and then sustain at those levels.

Many businesses zero in on the concept of Lean "Manufacturing," leaving out the systemic support actions that are vital to enable optimal performance. Admittedly, it is very hard not to get stuck on that because systems are fine-tuned to do manage inventory and direct labor costs. Once that's optimized, it's very difficult for transactional business leaders to turn such a dispassionate and equally aggressive view on themselves. What's good for the goose should be good for the gander. Has anyone considered the types of processes that would be in place if the direct labor teams set about to evaluate and optimize transactional flows with as much vigor as the front office team evaluates operational flows?

Either way, the next best step is to focus on transactionals. Those manufacturing systems that were just optimized were previously designed and delivered in a non-optimized manner by the transactional systems. Imagine the cost avoidance and margin improvement potential had those processes been optimized prior to release (This is where DFSS or DFM comes into play). If the front end is not managed, the business will continue to shoot itself in the foot as future processes are designed and implemented non-optimally.

So, the overarching Lean theory is this: ***Lean is about effective <u>and</u> efficient flow.***

Everything flows, both things and data. If it doesn't flow right the first time, losses are incurred. Whether those losses are in product capacity, scrap/yield, rework, overtime, premium freight, excess and obsolescence, material shrink, time loss, etc., it all adds up. Not all of these things are easy to quantify, but they all do make a difference. In 1985, the Harvard Business Review issued an article, titled "The Hidden Factory" describing some of the not-so-soft losses incurred by businesses when things aren't being done right the first time in the general office environment. These losses can be staggering. Overhead labor is the highest form of loss in an organization when not managed effectively. While often necessary, these activities are truly not value added from the product-perspective. The customer only buys the finished good. As such, a relentless, introspective focus on improving the effectiveness and efficiency of these processes should provide a higher and permanent return on the business' investment of time (Better yet, develops the processes first, and then staff them – just like in operations).

So, let's say that an integrated Lean system is established, and the business achieves Lean nirvana, what is the reward? Those physical savings that were tangible and real at the outset (assuming there is no backsliding and the savings are sustained) are now a part of the business' standard cost structure. How much more can be done to get additional results? Is this why Lean might seem to be so short-lived after it is executed within some organizations? It also seems to just stop at the plant floor. There is, after all, a point of diminishing returns. So, besides the "up front" money, what else does an effective Lean culture "buy" for the business? How can the value of sustaining efficient systems be identified? Only by tracking the lagging losses incurred when those efficiencies are lost? What is Lean's long term value?

The value of Lean is realized through three phases of maturity (**Figure 9.1**):

Phase I. Rapid Cost Recovery. Execute "within the four Walls" of a site or locale.

Phase II. Overhead Recovery and Cost Refinement. Execute via internal sustainment and addressing transactional processes that impact "the four Walls" of the business.

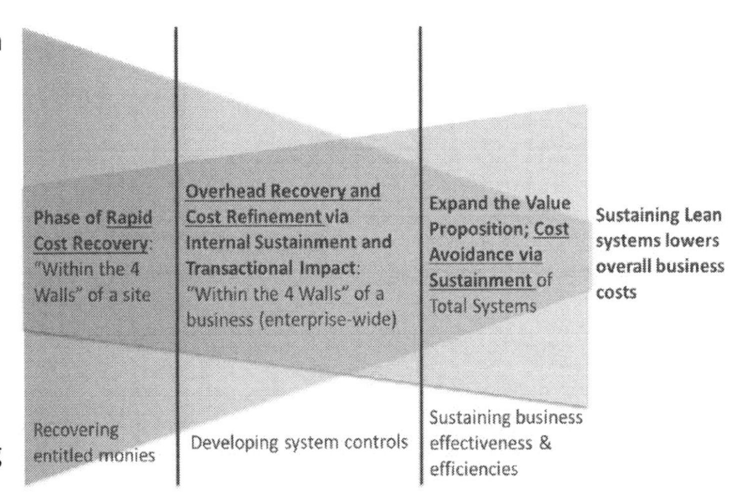

Figure 9.1

Phase III. Cost Avoidance Through Sustainment. Sustain and expand value propositions with external partners. Total cost of business to sustain should be less then losses incurred up front.

Note that several of the graphs in this chapter are shown in earlier parts of the book. They are also included in this section for easy reference and to provide additional context.

Phase I: Just Starting

Monies seem to fall from the Lean "sky" when businesses initially begin their Lean journeys. It can be "self-funding" if an organization is using a consultant to help them get started (The monies recovered can pay for the consultants; now there's a return). Working capital recovery from materials management via value stream mapping and execution of kanban can be extreme and immediate. A few key metrics are LT and enterprise-wide DIOH (**Figure 9.2**). If the business is able to quantify what one day of DIOH is worth, this can be rapidly translated into cash. The DIOH cycles below demonstrate both bridge builds for plant transitions and planned maintenance shut downs. As the PFEP and TTR are leveraged, only the right quantities of each item are manufactured, and only when required. Enterprise **Premium Freight** expense as a percent of COGS (**Figure 9.3**) as well as E&O (**Figure 9.4**) can improve rapidly. An unexpected bonus that our teams experienced was the ~34 percent reduction of its various forms of "**Rolling Stock**" (forklifts, golf carts, etc.) as the process flows were optimized (**Figure 9.5**).

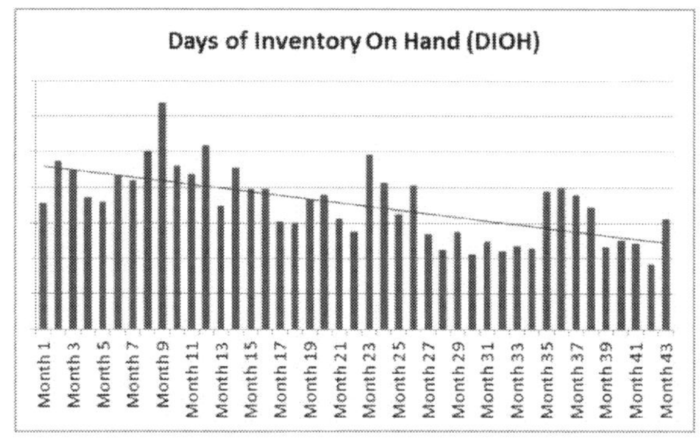

Figure 9.2

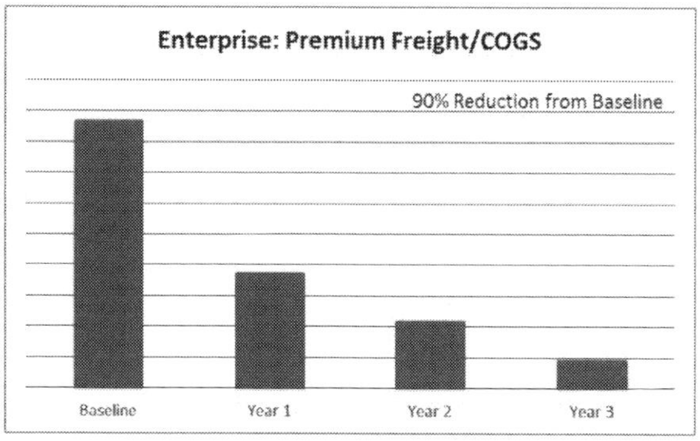

Enterprise: Premium Freight/COGS

90% Reduction from Baseline

Baseline Year 1 Year 2 Year 3

Figure 9.3

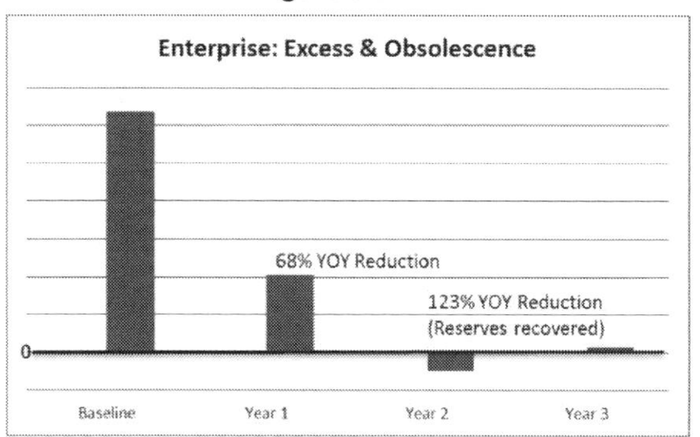

Figure 9.4

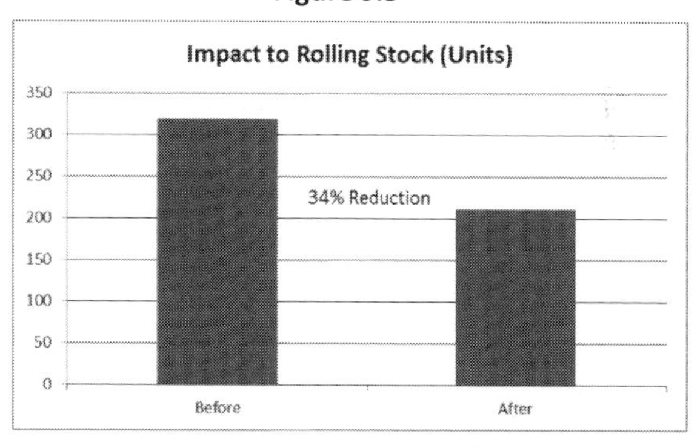

Figure 9.5

Flow cannot happen if the quality (or effectiveness) of the process is not correct. This is not only about "scrap" (excluding financial inferences about scrap having "value"), it's about the process not being done right the first time and the additional losses incurred with its management and recovery; **Scrap (w/o recovery) as a percent of COGS (Figure 9.6)**, Rework*, **OEE**, **Productivity** or **Throughput** and **Shrink (Figure 9.7)** are example measures in this area.

***Note:** Accuride products are largely non-reworkable; hence there are no trend metrics for this common KPI.

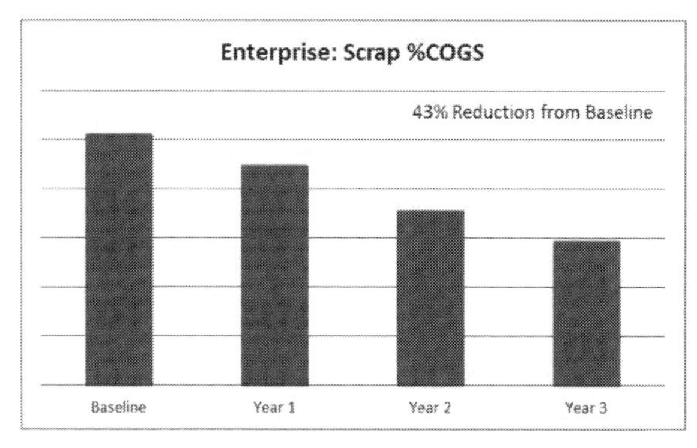

Figure 9.6

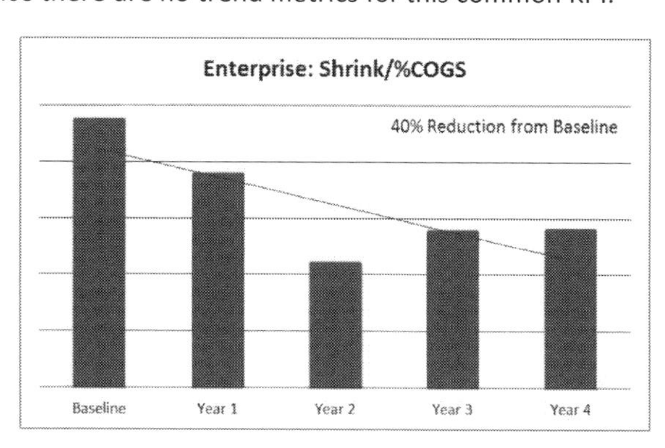

Figure 9.7

A side message on the notion of scrap having "value:" If the accounting team is going to account for the recovery "value" of product scrap, they should also include the additional losses with respect to double/triple handling and storage of the scrap, secondary capacity loss to cover the remake of the initially lost product, inspection, overtime, premium freight, etc. A full value of the cost of that lost unit should be considered. In truth, there may not be much value to the business in performing this level of analysis. A more serious effect is the impact of market fluctuations on the price of scrap (**Figure 9.8**). This can often have an inverse effect on "yield performance." Higher value scrap gives the appearance that the process is experiencing a slower rate of loss while lower valued scrap makes it appear that a higher rate of loss is being generated. This muddies the evaluation of process effectiveness with respect to the internal handling and process losses being incurred. The bottom line is that the process should be honestly evaluated as to what it is *yielding from a unit or "component" perspective* for focused improvement to happen.

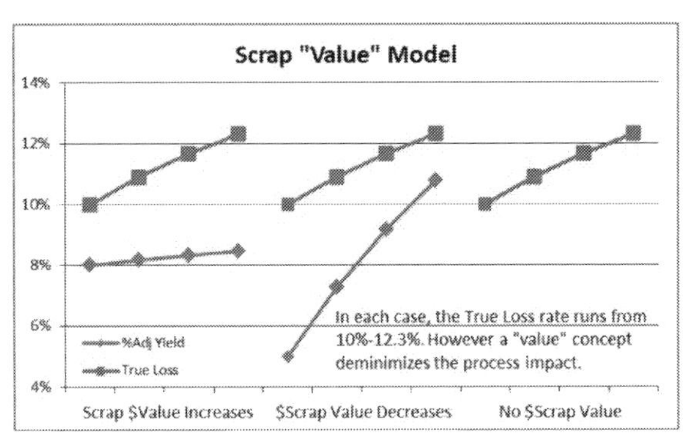

Figure 9.8

5S "findings" can generate cash to sell unneeded items and/or significantly delay re-ordering of "found" materials in the various caches located throughout the organization. In some cases, sites may not have had to reorder the majority of their office supplies for years. Findings have included gauges, tooling, fixture components, office furniture, raw material and unique surprises including money stashes and kittens. **MRO as a percent of COGS** is a metric to consider.

Lean impact monies can also be recovered in several forms during the implementation of standard work. While an evaluation and rebalance of direct labor is critical largely due to the sheer volume of personnel, indirect labor and overhead balancing (**Figure 9.9**) can also contribute significantly to the business' bottom line. **Figure 9.10** shows the change in percentage of revenue per employee from the original baseline (This chart currently excludes Accuride's most recent acquisition in Europe). In an organization, a rigorous evaluation of all types of work that are performed helps to reduce waste. All forms of work content should be evaluated:

1. Those manning machines for machine balance evaluation.
2. Direct labor for non-machine based activities where value added work is performed.
3. Leader Standard Work (LSW) for team leaders and supervisors.
4. Indirect labor content, such as materials handling, maintenance, etc.

5. Overhead labor content where repetitive work is conducted, such as call centers, debit memo processing, employee hiring, etc. See **Figure 9.11** for a corporate finance team sample.

Accuride teams started with the direct labor content and ***conducted at least "2.5" iterations across all of its identified workflows over five years*** (Newly acquired sites are in their first iteration while mature sites are in their third). Natural attrition enabled the majority of those impacted to be positively redirected into other areas of the business. As we've seen the opportunities in the "back end" of the factories normalize, the Accuride team recognizes that these inefficiencies were "built-into" these work flows in the first place as a result of legacy launch processes or through the incomplete resolution and/or verification of quality system issues. Work balance and material flow is now part of Accuride's new product development process called AccuLaunch™. While there may be some refinement after launch, we do not anticipate the significant levels of waste to be re-incorporated into future releases of new products that rob us of future margin and business agility.

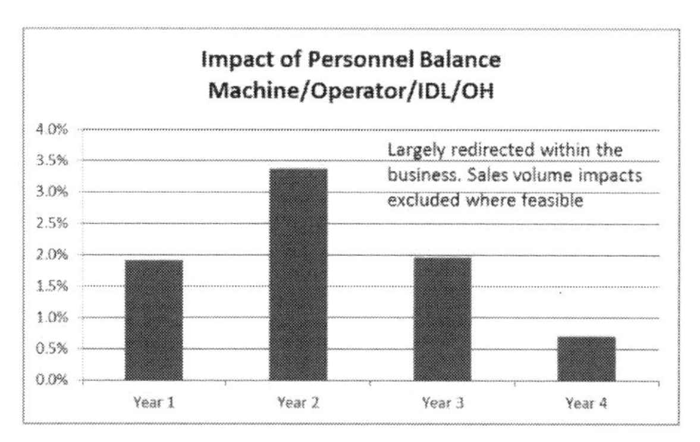

Figure 9.9

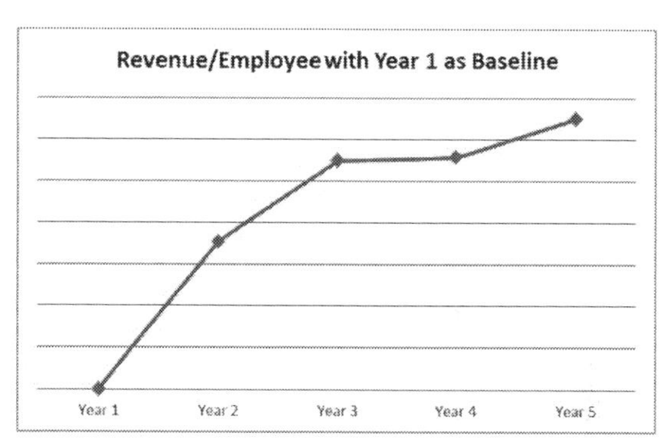

Figure 9.10

At an AME conference, a speaker from OC Tanner advised that it had well recognized its excess staff condition early on it its journey. The gentleman looked right at the audience and stated that they did not involuntarily impact a single individual due to their Lean journey. They let natural attrition absorb each gain. The best-of-the-best Lean companies strive to manage by this philosophy. This is where true employee loyalty, engagement and long-term systems sustainment comes from.

Accuride's history of involuntary personnel changes has largely been due to market changes in sales volume. While not fully successful in modeling OC Tanner's benchmark approach, we strive to develop inclusive processes enabling people to feel as safe as possible so they can proactively engage to support these efforts.

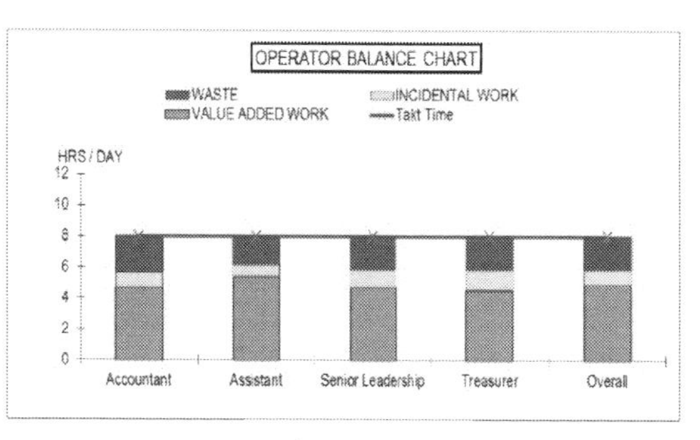

Figure 9.11

If we were not as effective as we are because of our Lean systems, the business would likely not be here today.

Phase II: Internal Sustainment

As gross forms of waste are managed across the "individual" business sites, the focus moves upstream to evaluating and creating transactional flows that are effective and efficient. It is more costly for an organization to regress and lose its Lean systems. Not only do the original inefficiencies and losses return, but repeated recovery is often not effective. Management of operations can only be as good as the information that it receives. After the "four walls of a site" are optimized, the focus broadens outward to the "four walls of the business." We call this the "inside out" approach. This includes the evaluation of the Enterprise's ability to:

- Deliver optimized products and processes where the sites do not incur long-term losses in scrap and inefficiencies. A tailored launch system with DFSS/DFM processes better prevents the release of non-capable products at inefficient production levels. QCD and additional features incorporating ergonomics, labor balance and material handling flows are pre-planned. This takes an intensive up-front effort with a well-trained cross-functional team. An AccuLaunch™ QCD KPI sample is modeled in **Figure 9.12**.

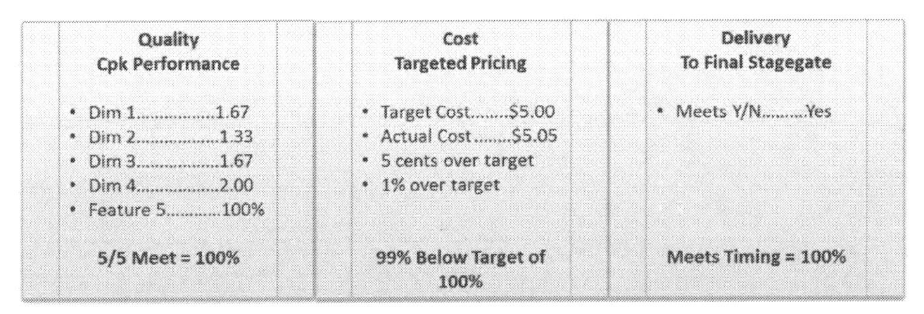

Figure 9.12

Assuming these results hit the trifecta, back-end margin eroders will not be incurred during product/process releases for any significant length of time.

- Concurrently conduct an intensive review of all products that are not meeting their cost objectives. These products should be evaluated and rationalized accordingly. Lessons learned from these evaluations are fed back into the AccuLaunch™ process to prevent recurrence.
- Determine from where to optimally initiate orders via the E-VSM process. This "inside-out" approach enables further working capital (specifically in raw material and WIP) and **Lead Time** reductions across the enterprise (**Figure 9.13**).
- Further improve the quality of vertically integrated products that are shipped to each other (Intercompany Parts Per Million or I/C PPM). The "Hidden Factory" losses incurred in

shipping, sorting and reproducing products sent I/C can be significant. See **Figure 9.14** for a sample of I/C PPM impact for one sub-component delivered solely within the company.

- Further leverage supply chain synergies via raw materials (focusing on QCD), MRO supplies and Vendor Managed Inventories (VMI) analysis. This also assists the supplier efforts to reduce inventory and improve quality in order to reduce costs (**Figure 9.15**).
- Ensure a transactional scope beyond production; management should enable a continued focus on business support functions, such as, but not limited to:

  - **Human Resource Processes**: Time to fill positions, effectiveness (**turnover**) of the new hires, employee engagement, utilization and effectiveness of benefits, etc.
  - **Finance Processes**: Month end close, third party audit findings/timing, debit and credit memo management, journal entries, internal and external reporting, etc.
  - **Legal Processes**: Patent processing, document management, etc.
  - **Corporate Processes**: Strategic/business planning, merger & acquisition processes, etc.
  - **IT Processes**: Reports, data management, etc.
  - **Sales and Marketing Processes**: VOC systems, surveys, sales processes, etc.

Several examples are shown in Chapter 4. There are many external references that focus on how to optimize these transactional process flows. These are often overlooked areas of opportunity. Again, these systems largely drive and support operations. Management of operations can only be as optimized as the systems that feed into it.

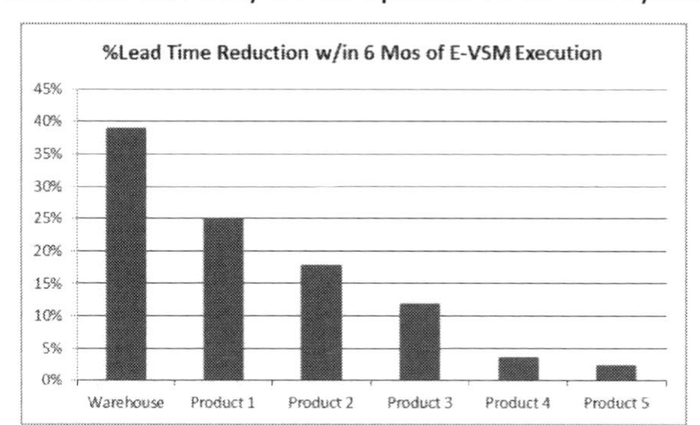

Figure 9.13

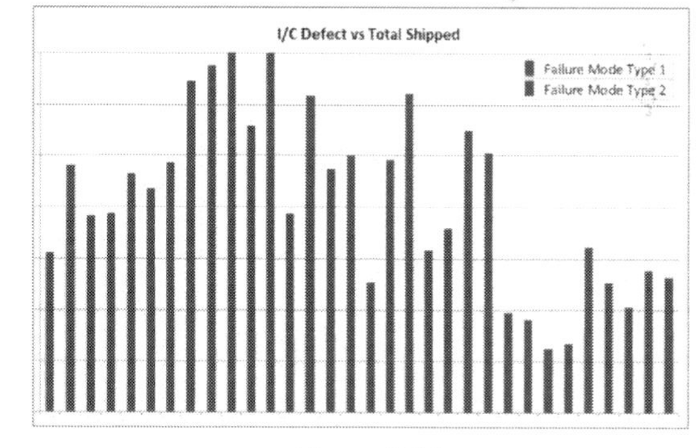

Figure 9.14

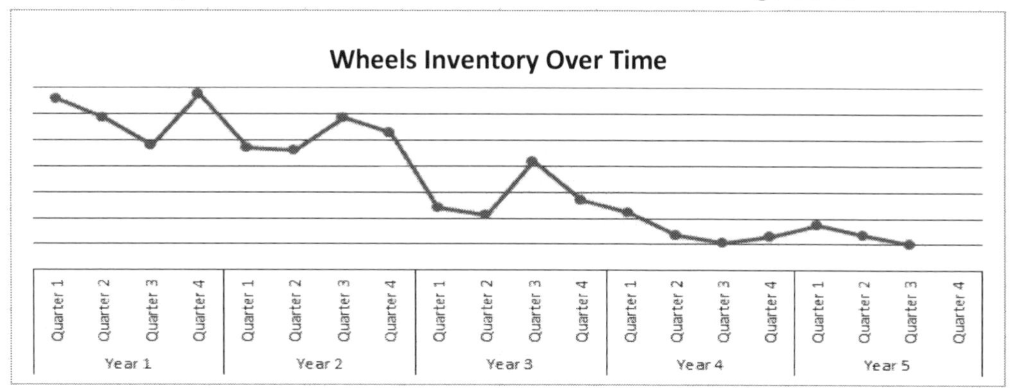

Figure 9.15

As scrutiny is placed upstream onto these transactional areas, the same opportunities are identified to redeploy some of these personnel as their scope of work changes. Getting "volunteers" to opt out happens naturally. Some people will just leave and/or seek other opportunities. However, if that doesn't happen, a catch-22 emerges where people just won't support the improvements if their livelihood is at risk.

Another organization, a Fortune 500 company, shared their story where they had developed a "Swing Team" of overhead personnel displaced due to Lean efforts. The team varied in size from four to eight employees. As attrition occurred, this team was used to fill those roles. There were no hires for three years; the Swing Team was used as an incubator. The savings and process improvements garnered by that self-assured team were legendary across the business. The personnel were in high demand and promoted into other parts of the business as natural opportunities arose. Due to the nature of the aggressive improvement process, others continued to populate the swing team as the incumbents filled attrition positions. When the business unit leader was tapped for other things, a new SVP was introduced. The SVP was shown the record process levels that the business had been achieving with respect to Sales/Employee and various other process efficiencies. The SVP looked at the Swing Team and said, "Hmm, these people don't have formal roles." After a quick whack to eliminate these "excess" positions, the SVP was largely lauded for the short-term gains and was later promoted; back at the ranch, transactional Lean was effectively halted, not only in the business, but also across the organization. That act killed the culture of improvement and engagement that had been flourishing. It did indeed bring a rapid ~$500K to the bottom-line; however, this leader was unable to recognize the longer-term business value that had been sacrificed.

Phase III: External Partnering/Engaged Culture

The early successes of recovering the operations processes should now have largely been absorbed back into the cost structure. The optimization of transactional processes should also now be mitigating and/or preventing the release of issues that were previously embedded in the business systems. Assuming the teams are now able to sustain these levels and continue on with iterative cycles of improvement, the remaining frontier is that of partnering as *assertively* as possible with the organization's external partners: Suppliers, Customers and local communities.

As a customer, the organization should have some control or leverage with helping its supply chain to optimize these incoming dynamics (See Chapter 8). Remember that suppliers may have key interactions at both the front and back end of the business with respect to product/process delivery. From the front-end side, optimizing the pull process with established market sizes along with delivery methods help to reduce working capital needs. Shared engineering development, which optimizes product effectiveness, is another common area of partnership.

Some recent examples that Accuride has been a part of includes the development of a series of world-class steel wheel coatings, such as Steel Armor™ (**Figure 9.16**) and EverSteel™ (**Figure 9.17**). Via the AccuLaunch™ process and intensive supplier partnering, Accuride's steel wheels are revolutionary (no pun intended) with respect to their longevity. Glance over at the wheels of the next 18-Wheeler that you pass on the freeway. How rusty are they? These coatings will help to prevent the onset of rust and enable a longer lasting product on the road.

Figure 9.16

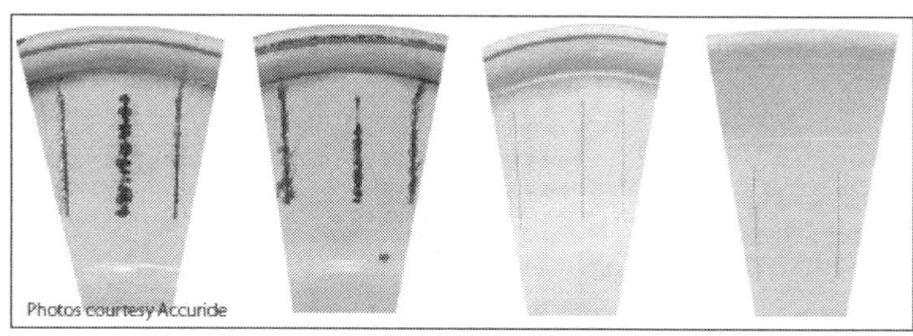

Left to Right: Accuride Competitor 1 Test Sample; Accuride Competitor 2 Test Sample; Accuride Steel Armor Test Sample; Accuride EverSteel Test Sample. Cyclic corrosion testing results (1) and (2) competitor steel wheels (3) Steel Armor® wheel and (4) the EverSteel® wheel. The new coating process increases lifetime from 20 cycles to more than 125 cycles.

Figure 9.17

From a supplier perspective, the organization may not find its customers as willing to engage in Lean systems. Some of them have not yet discovered the full benefits of these systems. For the ones that Accuride has been able to partner with, significant inroads have been made in providing value added services; adding intrinsic value that encompasses more than just delivering a part. Cross-developing pull markets has reduced FG inventory for both organizations *while sustaining throughput*. Leveraging team detailed OTD data has enabled sites to dramatically improve Customer Requested OTD (CRD) by sharing effectiveness results of third party carriers. One strong area of success has been managing the on-time pick-up and delivery of customer directed carriers. While Accuride sites largely enjoy a 100 percent OTD for MADD, this image can be heavily tarnished when the dock and no one shows up to pick up the product. Of course, the site gets tagged with the non-delivery. By working through strong cross-functional teams that included Sales, Supply Chain and the Customer, almost all of these "miss-connects" have been resolved (**Figure 9.18**).

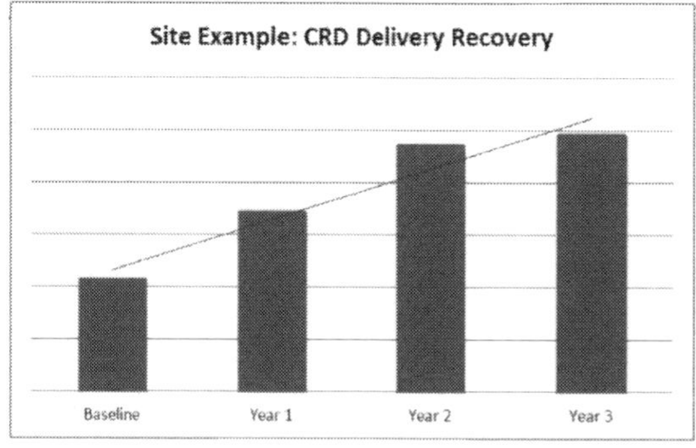

Figure 9.18

As the flows are optimized across "all of the processes," Lean enables a systemic form of customer focus: with both internal and external customers. It enables an organization to stay relevant and partnered with their customers. See Carlzon's comment in Figure 9.19 regarding customer focus.

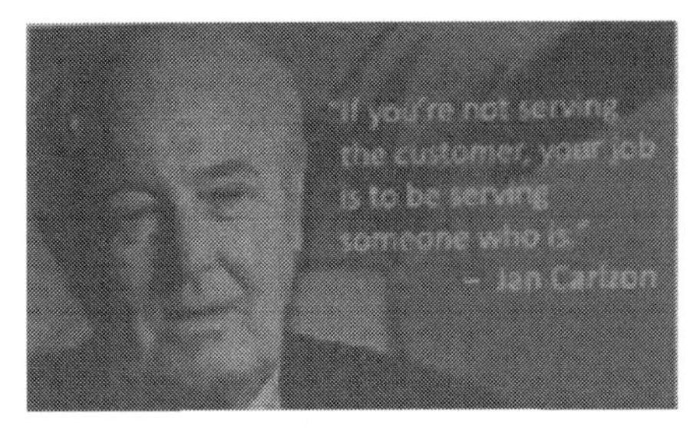

Figure 9.19

Effective and efficient flow systems should enable optimal competitiveness with respect to QCD. It is meant to enable a sustainment of the lowest practical cost levels while increasing business agility.

We wrote this book so that our experience may be used to help others on their journey. Our reflection across this process has continued to remind us that we are still on ours. We believe that other companies can leverage these ideas to see if they can achieve 50% reduction in their Lead Times so that they can enjoy a 50% increase in productivity and a 20% reduction in Cost Per Unit, enabling them to be more competitive. We believe that any business can use these Lean tools and systems to help them improve, grow and to become world-class in their industry.

We hope that you will find our story helpful and relevant to you, and best of luck on your 50-50-20™ Lean Journey!

Key Takeaways

1. Lean is about effective and efficient flow across product, process (internal site and across the enterprise) and partners (customers, suppliers and third parties).
2. Many stop at product ("Lean Manufacturing") because they may not recognize the systemic avoidance benefits of following through with processes and partners.
3. There are three phases of Lean management: Rapid Cost Recovery, Overhead Recovery and Optimization and Cost Avoidance by Sustainment.
4. Lean management is a continual journey.

Appendix A: Process Capability Index

When a process follows a Normal Distribution, shown in **Figure A.1**, a property of this distribution that we should expect is that a defined percentage of the area around the mean will be covered by the value of the distance from the mean.

To keep it simple, the area for 1, 2 and 3 standard deviations on both sides of the mean are given in **Figure A.2**.

This means if our specifications are at the limit of ± 3 Standard Deviations, we can expect 0.27 percent of the population to fall outside of this specification range as shown in **Figure A.3**.

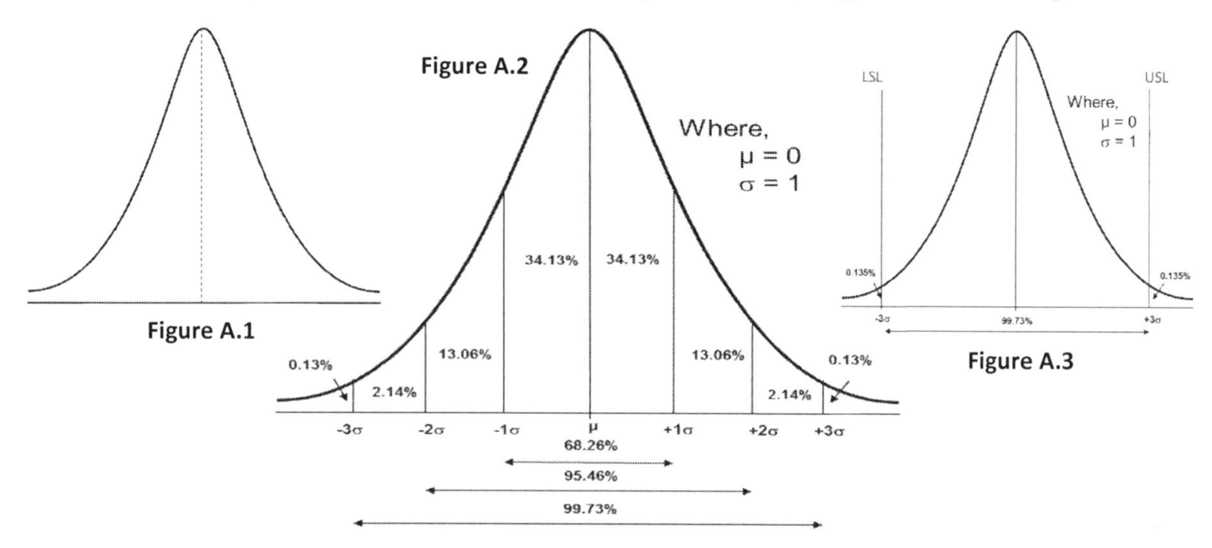

Figure A.1

Figure A.2

Figure A.3

In order to meet the specifications more frequently, we would like to reduce the standard deviation. It is not necessary for the process to be centered all the time. Both the issues of narrowing and shifting are addressed by the Process Capability (Cp) indices.

The width of the Normal Curve is represented by six standard deviations. Cp measures the process capability by comparing the ratio of voice of customer, represented by the specification range and the voice of the process, represented by six standard deviations as below:

Cp = (USL-LSL)/Six Standard Deviations

In other words, Cp tells how many bells can be fitted in the specification range as can be seen in **Figure A.4**.

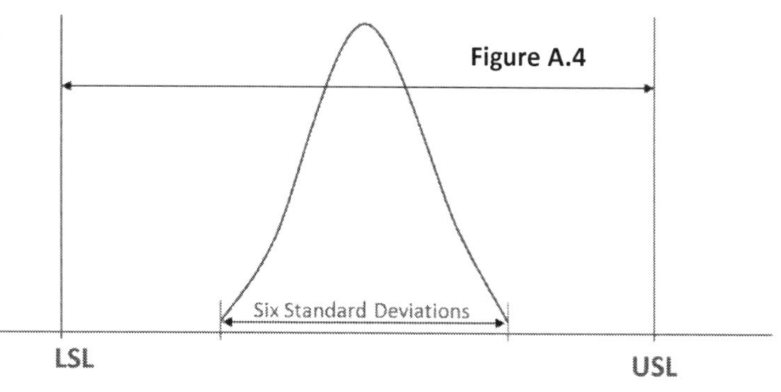

Figure A.4

In reality, however, the process is not always centered and may move closer to the upper or lower specification limit. Intuitively, we know the risk is higher when the process is moving closer to one side versus the other. This is addressed by Cpk, which takes this shift into account. We now divide the full bell into two haves and compare their fit as shown in **Figure A.5**. While this example shows that the shift is closer to the USL, the same logic would apply if the shift were closer to the LSL. Cpk is defined as the ***minimum*** of either:

Cpk_U = (USL-Average)/(Three Standard Deviations)
Cpk_L = (Average-LSL)/(Three Standard Deviations)

Figure A.5

Cpk quantifies the risk of being closer to either specification limit. The maximum value of Cpk is equal to Cp.

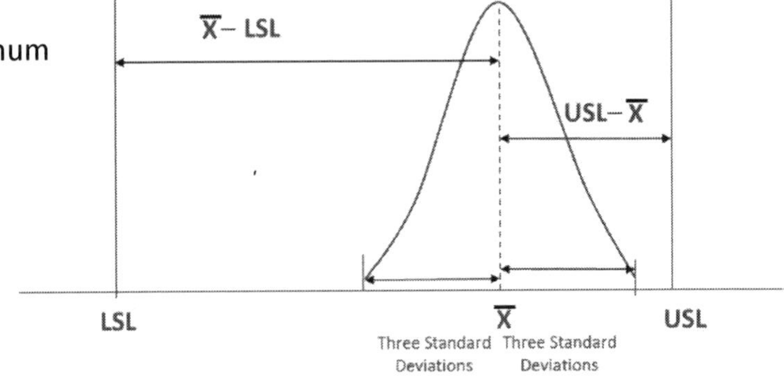

Biographies

Jeanenne "Jd" Marhevko (yes, the "d" is lowercase)

Jd Marhevko is currently the VP of Quality, Lean and EHS for Accuride Corporation. Four Accuride sites have won the prestigious Manufacturing Excellence Award (in 2014-2016). She has been involved in Operations and Lean/Six Sigma efforts for almost 30 years across a variety of industries. Jd is an ASQ Fellow. In 2016, Jd was awarded both the Shainin Medal and honored as one of the top 100 Women in Manufacturing by Washington D.C.'s Manufacturing Institute. With ASQ she also holds the CMQ/OE, CQE, CSSBB and is an ASQ Certified Trainer for a variety of Quality Tools. She is an MBB and has held various Baldrige Assessor roles. Jd is a Past-Chair of the ASQ Quality Management Division (QMD), a 24,000 member organization and supports several ASQ divisions in a variety of capacities. Jd is a Howard Jones awardee, the QMD's highest honor. She holds a BSE from Oakland University in MI and an MSA from Central Michigan University.

Arvind Srivastava

Arvind Srivastava is a Corporate MBB/IE at Accuride Corporation. Prior to joining Accuride, he worked as a process improvement consultant and supply chain professional at multiple organizations across various industries in the US and India. Arvind is an adjunct faculty at Illinois Institute of Technology (IIT), Chicago and co-author of Stat-Free Six Sigma book. He is a member of U.S. TAG to ISO/TC 176 (responsible for ISO 9001 standard) and U.S. TAG to ISO/PC 302 (responsible for ISO 19011 guidelines). Arvind is the Director of ASQ Chicago Section and teaches at its Training Institute. He holds BS (ME) from IIT Roorkee (India), MS (IE) from NITIE Mumbai (India), and MBA from Benedictine University, Lisle, IL. Arvind is a certified Six Sigma Master Black Belt (Accelper Consulting), Master Business Innovator (IIT) and Quality Auditor (ASQ). He is passionate about improvements in effectiveness and efficiency; reading and spreading knowledge; and old Hindi songs.

Mary Blair

Mary E. Blair is the Senior Vice President of Supply Chain and has been in this role since October 2011. Ms. Blair is responsible for managing Accuride's global supply chain initiatives and leads a team of seven disciplines. Previously, Ms. Blair was Vice President of Global Procurement at United Components, Inc. from 2008 until joining Accuride in 2011. Prior to UCI, Ms. Blair served as Director Global Sourcing for International Truck and Engine Corporation (Navistar) from 2006 to 2008. From 1984 to 2006, Ms. Blair had a distinguished career with General Motors Corporation. For GM, she served in various leadership roles in Worldwide Purchasing and Supply Chain, including most recently as Global Director Chemical Commodities. Ms. Blair also was named Professional Fellow for her advancement in technical expertise within the chemical commodity area. Ms. Blair is a graduate of Ferris State University in Michigan and earned an M.B.A. from Central Michigan University. She is currently pursuing her Ph.D. from Capella University in Business, specializing in global leadership and has achieved Ph.D., ABD. Ms. Blair serves on the board of directors of the University of Evansville Institute of Global Enterprise.

Bibliography

- AccurideCorp.com. Accuride Quality and Lean Management System (QLMS)
- AccurideCorp.com. Accuride Supply Chain Management (SCM) System
- Advanced Product and Quality Planning and Control Plan (APQP), Second edition. Automotive Industry Action Group (AIAG). Chrysler Corporation, Ford Motor Company and General Motors Corporation. 2008
- ASQ-QM.org. ASQ Quality Management Division (QMD)
- AT&T's Statistical Quality Control Handbook. 1985. AT&T.
- Bill Waddell, Lean Systems Audit (2010), Bill-Waddell.com
- Bill Waddell. Blog Post. Toyota's Eyes. 01/27/2016
- Business Deployment Vol. II: A Leaders' Guide for Going Beyond Lean Six Sigma and the Balance Scorecard. Forrest Breyfogle. Bridgeway Books. 2008
- Constraint vs. Bottleneck by Chris Hohmann. 05/06/2014. https://hohmannchris.wordpress.com/2014/05/06/constraint-vs-bottleneck/
- GembaAcademy.com. Gemba Academy
- Harvard Business Review, September 1985 Issue, The Hidden Factory by Jeffrey G. Miller and Thomas E. Vollmann
- Harvard Business Review. Positive Intelligence. Shawn Achor. January - February 2012 Issue
- How the Mighty Fall. Jim Collins. Harper Collins Publishers, Inc. 2009
- Institute of Management Accountants (IMA). IMAnet.org
- Juran Quality Control Handbook, McGraw-Hill Book Company
- Lean Accounting: What's It All About? Maskell, Brian and Baggesley, Bruce. Target Magazine, Volume 22, Number 1. 2005
- Lean Enterprise Institute (LEI) (www.lean.org), The Plan for Every Part (PFEP) by Chris Harris. 04/2014.
- Lean Thinking, James P. Womack and Daniel T. Jones, Simon & Schuster. 1996
- LNS WebEx/Interview, Improving Quality Costs by Improving Quality Processes, Jd Marhevko. 02/2014 on "Cost of Poor Execution (COPE) as a Percentage of Cost of Goods Sold (COGS)". February, 2014. http://blog.lnsresearch.com/blog/bid/194667/How-a-VP-of-Quality-Improves-Poor-Quality-Costs-Executive-Q-A
- LSS 6001 Lean Six Sigma Black Belt Training International Standard, Lean & Six Sigma World Organization, 1st Ed. 01/16/2016
- Making Materials Flow by Rick Harris, Chris Harris and Earl Wilson
- Mastering Lean Product Development: A Practical, Event-Driven Process for Maximizing Speed, Profits and Quality. Ronald Mascitelli. Technology Perspectives. 2011
- Practical Lean Accounting: A Proven System for Measuring and Managing the Lean Enterprise. Brian Maskell and Bruce Baggesley.
- Quality Management Forum. Spring 2015. Vol. 41. No 1. Chad and Debra Smith, The Importance of Flow and Why We Fail So Miserably at it

Bibliography cont.

- Quality Management Forum. Spring 2015. Vol. 41. No 1. Daniel Zrymiak, Understanding Governance Within Organizational Excellence and Management Systems
- Quality Management Forum. Spring 2015. Vol. 41. No 1. Gary Cokins, Fixing A Kite with a Broken String – The Balanced
- Quality Management Forum. Spring 2015. Vol. 41. No 1. Grace Duffy, How Group Decision Making Helps with Functional QMS Strategic Planning
- Quality Management Forum. Spring 2015. Vol. 41. No 1. Jd Marhevko, Preface to this Special Issue of the Quality Management Forum
- Rebecca Morgan, AME Author, When Excellence Is Not Enough, www.ame.org., 09/2016
- Satisfied Customers Tell Three Friends, Angry Customers Tell 3,000. Pete Black Shaw
- TBM Consulting Group, Practitioner Briefing on Enterprise-Wide Value Stream Mapping: Create a Vision of Your Company That Really Puts Your Customers First. Jonathan Chong. 03/2013. http://www.tbmcg.in/about-tbm/index.html
- The Complete Guide to Mixed Model Line Design, Leone, Gerard and Rahn, Richard, Flow Publishing. 2014
- The Happiness Advantage. Shawn Achor. Crown Business. 2010
- Theory of Constraints, Eliyahu M. Goldratt, North River Press. 1990
- The Machine That Changed the World, Womack, Jones and Roos, Simon & Schuster. 1990
- www.amazon.com
- www.Wikipedia.com
- www.Yelp.com
- You Want It In A Nutshell? Here's the Lean Accounting Nut, Maskell, Brian. BMA Inc. 09/15/2015

Made in the USA
Middletown, DE
28 March 2019